ELASTIC BEAM CALCULATIONS HANDBOOK

JIH-JIANG CHYU

J.ROSS PUBLISHING

ISBN 978-1-60427-010-5

Printed and bound in the U.S.A. Printed on acid-free paper
10 9 8 7 6 5 4 3 2 1

Library of Congress Cataloging-in-Publication Data

Chyu, Jih-Jiang.
    Elastic beam calculations handbook / by Jih-Jiang Chyu.
        p. cm.
    Includes index.
    ISBN 978-1-60427-010-5 (hardcover : alk. paper)
    1. Elastic analysis (Engineering). 2. Girders—Mathematical models. 3.
Engineering mathematics—Formulae. I. Title.
    TA653.C495 2009
    624.1'7723—dc22

                                                      2009010289

    Direct all inquiries to J. Ross Publishing, Inc., 5765 N. Andrews Way, Fort Lauderdale, Florida 33309.

Phone: (954) 727-9333
Fax: (561) 892-0700
Web: www.jrosspub.com

TABLE OF CONTENTS

FOREWORD

I am delighted to recommend the book entitled *Elastic Beam Calculations Handbook* for engineering instructors, students, and engineers alike. This handbook is a very useful addition to the arsenal of engineering education tools, to civil and mechanical engineering in particular. I believe that it fills a gap and helps students learn critical thinking in the age of information technology. The book covers an important area of analytical competence in the form of formulation of elastic beam problems and closed form solutions. This will surely help students acquire the critical analytical thinking which has sadly been not adequate in the age of Internet search for knowledge.

An engineering practitioner concerned directly with structural analysis of elastic beam problems will find many examples of useful and neat solutions and the rationale for their validity in the standard terminology of structural engineering. The author has brought together in a unified manner so much of the beam solutions that until now were made available only to a few in engineering practice who, like Dr. Jih-Jiang Chyu, contributed fruitful ideas and techniques for obtaining comprehensive solutions.

W.F. Chen
Member of the U.S. National Academy of Engineering
Member of Academia Sinica in Taiwan
Honolulu, Hawaii

PREFACE

As a comprehensive analytic treatment on elastic beam problems, with balanced emphasis on both the theoretical and the practical, this book is a vastly expanded version of the author's *Goldenbrook's Little Red Book* (2004) both in spirit and in style and with the same approach I call *open-mindedness*. The previous book was written primarily for students. The prevailing trend in education advocates critical thinking and promotes continuing education, as exemplified by the requirements for Professional Engineer licensing. Therefore, this book is intended for students and their teachers, as well as all structural engineers and applied mathematics professionals.

The present work entitled *Elastic Beam Calculations Handbook* is a resource book with insights to answer in sufficient detail the pivotal questions "when," "what," "how," and "why." Furthermore, because of the content, as well as the manner of treatment, which uses a rigorous deductive approach (from general to special cases), this book can effectively supplement textbooks on mechanics of solids for engineering and can enrich applied mathematics curricula through applications in engineering. In either case, it is designed to help the reader achieve his or her optimal educational goals.

In view of the current educational curriculum and professional training, this kind of book is needed now more than ever, especially in the context of computer applications. In fact, because computers are popular and indispensable nowadays (and will continue to be, of course), an engineering book that takes a rigorous and yet user-friendly analytic approach to the endeavor of revealing the hidden physical significance of solutions to elastic beam problems is invaluable. This is because this approach provides guidelines for the effective planning and execution of numerical work involved in a complicated project. Moreover, it can expedite the successful completion of the otherwise formidable task of embarking on such a giant project.

Numerous important and interesting topics are included herein, each one treated in detail, with cross-references to related topics pointed out clearly in order to delineate interconnectedness among various topics.

The author hopes that the reader finds this handbook inspiring, interesting, and useful. Last, but not least, any comments on any part of this book are welcome; these comments will be acknowledged, reviewed, and incorporated in the second printing as appropriate.

ABOUT THE AUTHOR

Dr. Jih-Jiang Chyu is a registered professional engineer with about four decades of professional experience in both consulting engineering (encompassing design and research of bridges, buildings, and special structures) and college teaching in engineering and mathematics. He holds M.S. and Eng. Sc.D. degrees in Civil Engineering and Engineering Mechanics from Columbia University and is a Fellow of the American Society of Civil Engineers and a Member of New York Academy of Sciences.

He has been very active in ASCE programs. To name a few, he was committee chairman of the ASCE MET Section 75th Anniversary Yearbook, host of Dr. Mario Salvadori's Lecture in 1996 at ASCE Headquarters in New York City, and on the U.S. delegation to the World's Engineering Resource Joint Conference in England in 1975.

His publications include numerous technical papers on constitutive laws of engineering materials, structural dynamics, general theory of elasticity, and wave propagation. He invented and developed the Plane Strain Analogy concept in the theory of elasticity and promulgated the Method of Analogy in engineering mechanics and structural engineering, thereby discovering a new frontier in terms of not only subject areas but also methodology, to facilitate advancement in both theoretical and practical aspects of engineering.

Furthermore, recognizing the importance of "connectedness" in many contexts, Dr. Chyu is an ardent advocate for promoting interactions among engineering research, design, and construction, as well as the linkage between engineering education and practice, as exemplified by his technical lecture presented in 2002 at the

ASCE MET Section Structures Group, entitled "Engineering Research and Design: A New Continuum with Illustrating Examples from Structures and Mechanics."

Dr. Chyu does consulting structural engineering design and research, as well as applied mathematics and engineering mechanics research work.

ACKNOWLEDGMENTS

This book, developed during the course of my work in design and research as well as my college teaching in engineering and mathematics in which I promoted interactions among these facets, is made possible with the help of many wonderful people.

I am grateful to Dr. W.F. Chen for his encouragement and for writing the foreword.

It is of special meaning to thank my alma mater, Columbia University, where I earned my master's and doctorate degrees.

ASCE aided me in focusing on "interactions" as I was on the U.S. Delegation to the U.K. for the World's Engineering Resource Joint Conference (1975), the host for Dr. M. Salvadori's Lecture in 1996, and the chair of the MET Section 75th Anniversary Yearbook Committee. The Technical Lecture (2002) that I gave on the interactions between engineering research and design, with illustrative examples from mechanics and structures, prompted the writing of this book.

Global Graphics, Inc. assisted me in word processing, and J. Ross Publishing took care of all the other aspects related to publishing the present work.

Finally, this book would not have been completed if not for the love, caring, and support given to me by my beloved wife, Dr. Chi-Oy Wei Chyu. On top of all the sacrifices and understanding on the part of a great spouse, she is a tremendous graduate school alumnus who inspires me with her expertise in mathematics education.

I would like very much to take this opportunity to thank them all.

Jih-Jiang Chyu
New York City

1

BASIC THEORY

This book uses innovative analytic approaches that combine tactful applications of mathematics with structural engineering, thereby helping the reader gain insight into the physical implications of the formulae presented. This means that an effective analytic treatment of the elastic beams will shed light on how the numerical work can best be planned and executed with clarity and optimal results, as well as a minimum of time, effort, and cost.

The writing philosophy of this book leads to a presentation at once both simple and logical, so that many important and interesting problems can be solved as corollaries of a general theorem. In this way, the reader will be able to see not only the trees but also the forest; this "big picture" approach is intended to be both enjoyable and inspirational.

It is not feasible to present all the important cases, and there is no need to make an effort to do that either, because by following the general approach presented here, the reader can easily extend the results by the principle of superposition. This is one of the reasons why we take the approach of starting with a general problem and later reducing it to simpler special cases. This approach is especially significant for multi-span beams.

Both solved problems with explanatory remarks and illustrative examples are presented throughout the book. For some problems, the actual solution process is purposely left to the reader in order to spur curiosity as well as to develop a capacity for exploration.

As its title implies, this book is a mathematically rigorous treatment of beam design problems based on elastic theory. Consequently, all the usual assumptions, principles, methods, procedures, etc. within the framework of linear elastic behavior of materials are either explicitly cited or implicitly utilized, in addition to the new approaches and methods developed and presented in this book.

For the sake of effective development of the subject matter, fundamental theory utilized in this book is delineated into two major parts: (1) universal guiding principles and (2) subject-specific approaches.

1.1. UNIVERSAL GUIDING PRINCIPLES

First, we will discuss the theory that is used throughout the book—the universal guiding principles. Here the key is to adhere to the utmost generality while performing mathematical derivations to the extent that is commensurable with feasibility as well as effectiveness and clarity of presentation. In this context, there are basically five aspects to be attended to: loads, the principle of superposition, materials and geometric properties, the solutions, and parametric investigations of the solutions. However, these are not individual entities separate from each other. In fact, they are interrelated, as will be seen later.

In principle, an applied load can be any load that is well defined in one way or another. One of the most useful and efficient ways to define a general load is by function of a real variable in mathematics, although it is by no means necessary to follow this approach. However, we will use the concept of mathematical function in our development for the simple reason that it is exact, convenient, and is, therefore, an efficient tool for achieving our goal of probing, with mathematical rigor, the structural behavior under any conceivable loading.

The most general load is expressed by a load intensity function. The independent variable here, depicted as the horizontal longitudinal beam axis x, indicates the location of the load intensity. For both theoretical and practical considerations, it is sufficient to use functions of a real variable that are piecewise continuous. We know that we can treat a piecewise continuous function as a string of continuous functions with a finite number of distinctive points of "jump." Thus, we can do our integration region by region, with one string corresponding to one region, by using a smooth and easy process. Moreover, the vast majority of beam problems of practical importance, and even most topics of theoretical interest for that matter, can lend themselves to treatment by means of elementary functions, which can be integrated with relative ease. Notable examples of such functions are polynomials and trigonometric functions. The coefficients of these functions can be any real numbers without any restrictions whatsoever. In other words, no matter how complicated a load case is, we can figure out a function tailored to the needs of the particular problem at hand. Once this is done, we are in business. That is the beauty of a general load intensity function.

In an effort to achieve the optimal scope of applicability of the results and at the same time to reveal the intriguing interrelationships among special cases, we will also apply the concept of generality to entities associated with a load intensity function.

These include, among other things, the extent of the loaded regions and the beginning and the end of such regions. As a result, a fully loaded span can be considered as a special case of a partially loaded span, and the solution to the (original) special problem is extremely simple once the general problem is solved.

There are two basic approaches to the general load problems. One is to treat them strictly as such from beginning to end by using integration during the course of the solution-seeking procedure. The other is to make use of the solution results for the problem of a concentrated force, henceforth called the generic problem for the obvious reasons demonstrated later, and obtain the desired solution through integrating the solution to the generic problem.

Certain important problems frequently encountered in practice can be solved either through the general approach outlined above or by a more direct and problem-specific procedure. One such example is the case of uniform load over an arbitrary area. Thus, we see that there is room for flexibility in choosing a method for or an approach to a solution.

The principle of superposition provides a powerful and handy tool for obtaining solutions to new problems through solutions to old ones by considering (1) different loads in the same region for all problems at hand, (2) loads in different regions for the old problems, or (3) a combination of the two.

Moreover, the method of integration is also an efficient application of the principle of superposition for solving a given problem under a distributed load of any type, including, of course, the one specified by a general load intensity function.

The modulus of elasticity and section properties of a beam are constant within a span, but any one of them can vary from span to span. Also, span length is an interesting and important parameter to be used for probing the behavior of the beam in terms of reactions, bending moments, shear forces, and deflections. Other useful and important parameters that can and will be used in the analytical investigation include the various characteristics of the load intensity functions, the extent as well as the end points of the loaded area, material and section properties of the beam, etc. Any of these characteristics can be quite general and/or specific, thus yielding a great number of topics of interest. Here again, the concept of generality is fundamental to developing the procedure and results for a solution.

The solutions to beam problems are obtained by various methods for different problems and are usually mathematical expressions for reactions, bending moments, shear forces, and deflections. These expressions contain all the relevant information about the beam and its supports or foundation. Therefore, any change in any one of the entities involved in a solution formula will have an impact on the solution. If we can keep track of the effects that the variation of the chosen entity has on the resulting solution and can ascertain with mathematical precision and rigor, then we are on our way to the successful completion of performing an efficient parametric investigation.

When we decide which entity in the solution formula will be the parameter u for our study, we are treating this formula as a real-valued function $f(u)$ of one real variable u. Then we can perform the usual task of determining the conditions for $f(u)$ to be an increasing or decreasing function of u and the corresponding domain for u; the conditions for $f(u)$ to have maximum or minimum values, together with the associated values of u and $f(u)$; as well as the conditions of determining points of inflection and the corresponding values of u and $f(u)$. This will entail the application of first and second derivatives of $f(u)$ in general. The defining conditions just mentioned often tie many other entities in the original solution formula in a very interesting and physically meaningful way. This is where the whole undertaking of parametric investigation pays its handsome dividend. Besides their usefulness, the process and the end results of investigations are very interesting in their own right.

The procedure to carry out this plan is to start with the newly obtained solution, which is a formula that explicitly expresses a reaction, deflection, or whatever the solution is for, and focus on one entity at a time in a manner described above. Then we consider the solution as a function of the chosen parameter in order to perform a mathematical analysis of a real-valued function of a real variable by the standard methods of calculus.

Throughout the book, in addition to the use of a common method to solve as many problems as feasible, different methods are also used to solve some problems. The point is to show the diversity of methods available while expanding the reader's horizon.

1.2. SUBJECT-SPECIFIC APPROACHES

Chapter 2 is preparatory in nature and outlines in detail methods and procedures specific to a topic. One of its goals is to give the reader experience using the general approach and procedure in preparation for tackling more advanced problems in later chapters. Chapter 2 also includes a rich collection of deflection formulae not readily available in other books.

The general principles usually applied for simply supported beams are those from statics for obtaining solutions in terms of moments and shear forces. A handy and convenient method for obtaining the deflection formulae is the method of conjugate beams. This is not only for special load cases but also for much more complicated and general load cases. These statements also apply for cantilevers, which are discussed in Chapter 5.

Chapter 3 deals with two- and three-span continuous beams. The treatment of two-span continuous beams employs a complete display of the following four categories of results for a given type of load: (1) general span lengths, materials, and section properties; (2) equal span lengths; (3) constant materials and section properties for all spans; and (4) constant materials and geometric characteristics through all the spans. There are four cases under each category. For three-span continuous beams, a different approach is taken in order to demonstrate an alternative way to organize and present results.

For topics covered under continuous beams, the standard method of least work based on energy considerations is used first to obtain the redundant reaction(s) for the statically indeterminate structure. This is followed by the application of statics to obtain the remaining reactions. The remainder is the same as solving a problem in statics as far as member forces are concerned.

For beams on elastic foundations, solving ordinary differential equations of the fourth order is the starting point. The general solution is then applied to various specific problems via the given boundary conditions under the prescribed loads. This solution is usually the deflection at an arbitrary point on the beam. However, in some problems, bending moments are also sought. In either situation, we will begin an analytic study by varying one parameter at a time to see its effect on the resulting solution. This parameter is, in most cases, a factor of the independent variable of the deflection or the bending moment function. Note that the product of this factor (which itself typically is a rather complicated function of material characteristics of both the beam and the underlying foundation, as well as the section modulus of the beam) and the horizontal location of the beam section under consideration is the independent variable of the solution function. There are many ways to reach this given value of this chosen parameter; hence, we see the opportunity for computer-aided study regarding these characteristics. The "multi-level matrix approach," as outlined below, was originally devised for analytical study and is one of the many possible ways to tackle the project with ease and efficiency.

The "multi-level matrix approach," which is a special term coined here for convenience and lack of a better name, is employed extensively in Chapter 6 and to a certain extent in other chapters for the exploration of a given solution function for a reaction, deflection, or bending moments. First of all, this is not exactly mathematical matrix analysis; rather, it has, in essence, the spirit of the matrix and goes beyond that by allowing us to handle the matrix concept level by level, both philosophically and methodically, to achieve our goal.

The multi-level matrix approach starts off with the given solution. An example would be a reaction as a function of the parameter of our choice only. This function may be a polynomial, a fractional function, or a combination of the two. It also may be some other function. We will call this solution function the original function for convenience of use in the following.

The entire collection of entities in the original function consists of not only this parameter but also many other entities. Each of these entities is related to some others in one of many complicated ways. Furthermore, members of the whole class of these relationships interact with each other, displaying explicitly or implying implicitly the physical significance in their own right. These members appear as individual terms, coefficients, or functions themselves, all within the realm of the given original function, so we begin our parametric study by focusing on one main parameter at a time. Thus, in essence, we are dealing with families of functions, and sometimes relatively large ones. It is not surprising that we see families that encompass three generations, and that is why the approach is called "multi-level."

Each of these new functions contains several variables within the original function, which itself may be a coefficient in the original function R, and is a function F of only one real-valued variable when the rest of the entities in this new function F are held constant. We may choose one particular entity in F as the new parameter to study its effects on F. The first step is to discern whether F is unconditionally positive or negative. If that is the case, we will immediately assess its impacts on R regarding its functional form and related aspects. If not, we will see the significance of the sign of F from the results via precise specifications of conditions under which F will be positive, negative, or zero. Consideration will be given to all possibilities and each condition will be explored in its own right first, followed by the assessment of consequences for the satisfaction of each such condition, including further probing into the mathematical results obtained thus far as well as the timely interpretation of influences on R in terms of both functional form and physical significance. If any of these cases is not possible, we will prove it and give reasons delineating why not. Moreover, we will state what the effects of any one of these situations are on the function R regarding several important aspects.

Once we treat F as a subject of exploration, we can find F' and F'', the first and the second derivatives of F with respect to the chosen parameter, and use these to determine the extreme values and stationary values of F, if any, and the associated values of the parameter giving rise to these extreme values and stationary values of F. We will also assess the effects of each of these conditions on the function R and determine the corresponding physical significance.

We can take a similar approach to F using another entity as the parameter. This is for one member of the family of functions under the original function R. We can treat other members of the family R that are siblings of F as well as the function R itself in a similar manner. That sums up the "multi-level matrix approach" in a nutshell.

An alternative way to obtain solutions to problems resulting from two- and four-span continuous beams with symmetry is shown at the end of Chapter 6. This is of special interest both theoretically and practically. Finally, possible future extensions of the present work and more examples that show the applications of the principle of superposition are suggested.

1.3. MATHEMATICAL THEOREMS

We now turn our attention to something entirely different in nature, something purely mathematical: an important, useful lemma which will be cited frequently in Chapter 6 for solving interesting problems of both theoretical and practical significance.

Lemma

Consider a set of entities of real numbers defined by

$$E_j = \frac{e^j - a^j}{j}, \quad j = 2, 3, 4, 5 \tag{1.1}$$

with

$$e = a + b \tag{1.2}$$

$$a > 0 \quad \text{or} \quad a = 0 \tag{1.2A}$$

$$b > 0 \quad \text{and} \quad b \neq 0 \tag{1.2B}$$

where e is not equal to zero. Then, we have the following conclusions

$$E_j > aE_k \tag{1.3}$$

where

$$k = j - 1 \tag{1.4}$$

and

$$E_j = aE_k = 0 \quad \text{for } j = 2, 3, 4, 5 \text{ if } e = a \tag{1.5}$$

Proof

First, let us consider the most general category where the entities e and a are not equal by looking at each of the four values of j separately, as follows.

Case 1. j = 2. Formula (1.1) becomes

$$E_2 = \frac{e^2 - a^2}{2} = (e - a)\frac{e + a}{2} > (e - a)a \tag{1.6}$$

This is because

$$(e + a) > 2a \tag{1.7}$$

which is the result of

$$e > a \tag{1.8}$$

obtained from the given premise that e and a are not equal as expressed in (1.2), (1.2A), and (1.2B).

Note that $(e - a) = E_1$ appears in (1.6). Thus, Expression (1.6) says the following

$$E_2 > aE_1 \qquad (1.9)$$

which is Formula (1.3) for $j = 2$.

Case 2. j = 3. Formula (1.1) becomes

$$E_3 = \frac{e^3 - a^3}{3} \qquad (1.10)$$

Thus,

$$E_3 - aE_2 = \left(\frac{b}{6}\right)[2e^2 - a(e + a)] \qquad (1.11)$$

However, from the given premise that e and a are unequal as expressed in (1.2), (1.2A), and (1.2B), we have

$$e > a \qquad (1.12)$$

Therefore,

$$2e^2 > 2ae > a(e + a) \qquad (1.13)$$

Substituting (1.13) into (1.11), we obtain Formula (1.3) for $j = 3$.

Case 3. j = 4. Formula (1.1) becomes

$$E_4 = \frac{e^4 - a^4}{4} \qquad (1.14)$$

Thus,

$$E_4 - aE_3 = \left(\frac{b}{12}\right)[3e^3 - ae^2 - a^2e - a^3] \qquad (1.15)$$

Again, however, e and a are unequal via the given premise in (1.2), (1.2A), and (1.2B), so we have

$$e^3 > ae^2, \qquad e^3 > a^2e, \qquad e^3 > a^3 \qquad (1.16)$$

Thus, from (1.15) and (1.16) we obtain Formula (1.3) for $j = 4$.

Case 4. j = 5. Formula (1.1) becomes

$$E_5 = \frac{e^5 - a^5}{5} \qquad (1.17)$$

Hence,

$$E_5 - aE_4 = \left(\frac{b}{20}\right)[4e^4 - e^3a - e^2a^2 - ea^3 - a^4] \qquad (1.18)$$

Again, however, $e > a$ because of the given premise in (1.2), (1.2A), and (1.2B), so we have

$$e^4 > e^3a, \qquad e^4 > e^2a^2, \qquad e^4 > ea^3, \qquad e^4 > a^4 \qquad (1.19)$$

and from (1.19) we obtain

$$4e^4 > e^3a + e^2a^2 + ea^3 + a^4 \qquad (1.20)$$

Thus, from (1.18) and (1.20), we obtain Formula (1.3) for $j = 5$.

Case 5. e = a. We have, from Formula (1.1), the result

$$E_j = \frac{a^j - a^j}{j} = 0 \qquad (1.21)$$

and also

$$E_j = 0 \quad \text{and} \quad aE_k = 0 \qquad (1.22)$$

when $e = a$ is the case.

Therefore, the lemma is proved.

As mentioned above, the purpose of presenting this lemma in its present form and content is for its frequent application to solve problems in Chapter 6; as such, the domain of the subscript j in the entity E_j, tailored to suit the actual need there, is naturally very limited.

Note that, then, from a universal point of view, the lemma specified above is actually a special case of a more general theorem. We could have stated and proved this general theorem first and then let the subscript j take its value required at a particular instance of application. However, by doing so, not only is it inconvenient and indirect for practical applications, but also we would miss the sharp contrast between the special case and the general situation by deliberately displaying the special case before the general one. We know that this is out of the ordinary in that we usually stick to the practice of doing the most general case first and then reducing it to individual special cases when solving a problem that calls for the intense treatment of many intricate cases which are, in most instances, even under several major categories.

Other reasons for doing what we are doing now are to arouse the reader's interest by changing the order of presentation (when that does not mean losing the charm of seeing the whole picture when an unusual sequence of displaying things suddenly takes place) and to leave more room for the reader to ponder other ways to achieve the same goal. For example, the reader may want to figure out the feasibility of utilizing a different method for proving the general theorem. In fact, in this connection, in addition to the proof presented below, the reader might suggest that we could do a proof by the principle of mathematical induction. However, we will soon realize that it is not necessary to embark on such a monstrous task. Such a task is planned mainly for the subscript j to increase without bound, which we do not really need when we compare it with the easy and clean direct proof of the theorem given below.

General Theorem

Consider any real numbers E_j, a, e with the properties specified in Formulae (1.23–1.26) below:

$$E_j = \frac{e^j - a^j}{j} \tag{1.23}$$

$$a > 0 \quad \text{or} \quad a = 0 \tag{1.24}$$

$$b > 0 \quad \text{or} \quad b = 0 \tag{1.25}$$

$$e = a + b \tag{1.26}$$

where e is nonzero and j is any positive integer.

We have, in general, when e and a are unequal, the result

$$E_j > aE_k \tag{1.27}$$

$$j = k + 1 \tag{1.27A}$$

for all positive integers k without any restrictions, but only for positive integers j that satisfy (1.27A).

Additionally, when $e = a$, we have

$$E_j = 0 = E_k \tag{1.28}$$

for all positive integers j, k stipulated above.

Proof

Let us consider the general situation where e and a are unequal first. Then, from Formulae (1.24–1.26), where e is nonzero, we have immediately

$$b > 0 \tag{1.29}$$

which means, in conjunction with (1.26), that

$$e > a \tag{1.30}$$

Now consider A defined by

$$A = E_j - aE_k \tag{1.31}$$

Then, upon using (1.27A), we have

$$A = \left(\frac{b}{jk}\right)\left[k(e^k + G) - ja\left(\frac{G}{a}\right)\right] = \left(\frac{b}{jk}\right)[ke^k + (k - j)G]$$

$$= \left(\frac{b}{jk}\right)(ke^k - G) \tag{1.32}$$

where

$$G = e^{k-1}a + e^{k-2}a^2 + \ldots + a^k \tag{1.33}$$

Next

$$e^n > a^n \tag{1.34}$$

for any positive integer n. Therefore,

$$e^k > e^{k-1}a$$

$$e^k > e^{k-2}a^2$$

$$\vdots$$

$$e^k > a^k \tag{1.35}$$

Now, adding up the left- and right-hand sides of each of the formulae respectively in (1.35), we obtain

$$ke^k > G \tag{1.36}$$

Therefore,

$$A > 0 \tag{1.37}$$

This completes the proof for the general theorem.

Finally, for the special case where $e = a$, direct substitution of this defining condition into (1.23) will yield the desired result immediately. Q.E.D.

In summary, this chapter, as its title implies, paves the way for the development of topics presented in this book.

Remark Regarding the Notations

The symbol e has absolutely nothing to do with the base of natural logarithm; all the other symbols and notations utilized in this book are defined where they first appear in the appropriate context.

2

SIMPLE BEAMS: AN INTRODUCTION
TO THE GENERAL APPROACH

In conformance with the general approach to the procedure for dealing with beams of various degrees of complexity and for convenience and clarity of presentation, we will begin with simple beams and in particular with a single concentrated force, so that the reader can, in addition to easy access to a detailed account of certain parts of the solution that are usually omitted from other books, get a feel for the general approach and be well prepared for more complex topics.

2.1. A CONCENTRATED FORCE AT AN ARBITRARY POINT ON THE SPAN

General Case

The beam with its loading is shown in Figure 2.1.

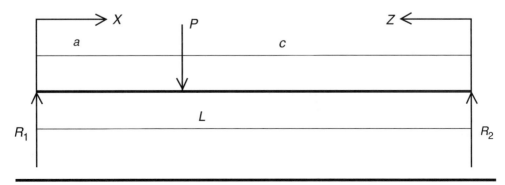

FIGURE 2.1. A concentrated force at an arbitrary point on a simple beam

The reactions are

$$R_1 = \frac{Pc}{L}, \qquad R_2 = \frac{Pa}{L} \tag{2.1}$$

The bending moments are

$$M = (R_1)x \quad \text{for } 0 < x < a \tag{2.2}$$

$$M = (R_2)z \quad \text{for } 0 < z < c \tag{2.3}$$

The maximum moment is $M_0 = Pac/L$, at $x = a$, the point of application of the concentrated force.

Note that in the above, a single horizontal axis x could be used throughout. However, for easier derivation and simpler presentation of results, especially for later chapters, the use of two horizontal axes is warranted. Now, for the sake of consistency, we will be using this approach from the onset, even though the advantage of doing so for a simple problem like the present one is not so obvious.

In the following, we will denote the modulus of elasticity by E and the moment of inertia of the cross section of a beam with respect to the horizontal principal axis by I.

The deflections are

$$Y = \frac{1}{EI}(M_0)\frac{x}{6}\left[3C - \frac{x^2}{a}\right] \quad \text{for } 0 < x < a \tag{2.4}$$

$$Y = \frac{1}{EI}(M_0)\frac{z}{6}\left[3D - \frac{z^2}{c}\right] \quad \text{for } 0 < z < c \tag{2.5}$$

where

$$C = \frac{L + c}{3}, \qquad D = L - C \tag{2.6}$$

The maximum deflection is

$$\max Y = \frac{1}{3EI}[aM_0^2 C^3]^{\frac{1}{2}} \quad \text{when } a > c \tag{2.7}$$

and is

$$\max Y = \frac{1}{3EI}[cM_0^2 D^3]^{\frac{1}{2}} \quad \text{when } c > a \tag{2.8}$$

The location of the maximum deflection corresponding to (2.7) is

$$x = (aC)^{\frac{1}{2}} \tag{2.9}$$

The location of the maximum deflection corresponding to (2.8) is

$$z = (cD)^{1/2} \tag{2.10}$$

Special Case. $a = c = L/2$

The reactions are

$$R_1 = R_2 = \frac{P}{2} \tag{2.11}$$

The bending moments are

$$M = (R_1)x \quad \text{when } 0 < x < a \tag{2.12}$$

The maximum bending moment is

$$M_0 = \frac{PL}{4} \quad \text{at } x = a = \frac{L}{2} \tag{2.13}$$

The deflections are given by (2.4), with

$$C = \frac{L}{2} \tag{2.14}$$

The maximum deflection is

$$\max Y = \frac{1}{12EI} (M_0)L^2 \tag{2.15}$$

or equivalently

$$\max Y = \frac{1}{48EI} PL^3 \tag{2.16}$$

located at $x = L/2$.

2.2. UNIFORM LOAD

General Case

The beam with its loading is shown in Figure 2.2. The reactions are

$$R_1 = wb \, \frac{c + \dfrac{b}{2}}{L} \tag{2.17}$$

$$R_2 = wb \, \frac{a + \dfrac{b}{2}}{L} \tag{2.18}$$

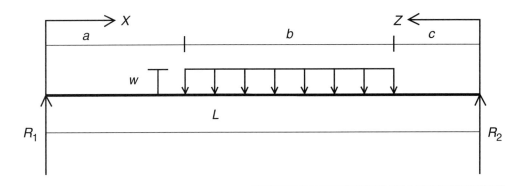

FIGURE 2.2. Uniform load on a simple beam

The bending moments are

$$M = R_1 x \quad \text{for } 0 < x < a \tag{2.19}$$

$$M = (R_1)x - w\frac{(x-a)^2}{2} \quad \text{for } a < x < e, \text{ with } e = a + b \tag{2.20}$$

$$M = (R_2)z \quad \text{for } 0 < z < c \tag{2.21}$$

with R_1 and R_2 given by (2.17) and (2.18) respectively.

The deflections are

$$Y = \frac{1}{EI}\left[S_1 x - \left(\frac{1}{6}\right)(R_1)x^3 \right] \quad \text{for } 0 < x < a \tag{2.22}$$

$$Y = \frac{1}{EI}\left[S_1 x - \left(\frac{1}{6}\right)(R_1)x^3 + \left(\frac{1}{24}\right)w(x-a)^4 \right] \quad \text{for } a < x < e \tag{2.23}$$

$$Y = \frac{1}{EI}\left[S_2 z - \left(\frac{1}{6}\right)(R_2)z^3 \right] \quad \text{for } 0 < z < c \tag{2.24}$$

where

$$S_1 = C - \frac{D}{L} \tag{2.25}$$

$$S_2 = \frac{D}{L} \tag{2.26}$$

with

$$C = (R_1)\frac{e^2}{2} + (R_2)\frac{c^2}{2} - \frac{wb^3}{6} \tag{2.27}$$

and

$$D = (R_1)\frac{e^3}{3} + (R_2)c^2\frac{3L - 2c}{6} - \left(\frac{wb^3}{24}\right)(3b + 4a) \tag{2.28}$$

The maximum Y for the region $a < x < e$ can be obtained by solving a general third-degree algebraic equation for the purpose of investigating parametric characteristics before plunging everything into the ocean of numerical data.

The maximum Y for another region is

$$\max Y = \frac{1}{EI}\left[(S_1)(x_0) - \left(\frac{1}{6}\right)(R_1)(x_0)^3\right] \quad \text{for } 0 < x < a \tag{2.29}$$

with

$$x_0 = \left[\frac{6(S_1)}{(R_1)}\right]^{1/2} \tag{2.30}$$

Special Case 1. a = 0

The reactions are

$$R_1 = \frac{wb\left(L - \dfrac{b}{2}\right)}{L} \tag{2.31}$$

$$R_2 = \frac{wb^2}{2L} \tag{2.32}$$

The bending moments are

$$M = (R_1)x - \frac{wx^2}{2} \quad \text{for } 0 < x < b \tag{2.33}$$

$$M = (R_2)z \quad \text{for } 0 < z < c \tag{2.34}$$

The deflections are given by (2.23) for $0 < x < e$ with $a = 0$ and are given by (2.24) for $0 < z < c$, with S_1 and S_2 given by (2.25) and (2.26) respectively and R_1 and R_2 by (2.31) and (2.32) respectively.

However, here we have

$$C = (R_1)\frac{b^2}{2} + (R_2)c^2 - \frac{wb^3}{6} \tag{2.35}$$

$$D = (R_1)\frac{b^3}{3} + (R_2)c^2\frac{3L - 2c}{6} - \frac{wb^4}{8} \tag{2.36}$$

Note that we can let $c = 0$ in the general case to obtain the same result for the special case if we prefer to have a simpler C, D and a somewhat more involved (2.23) which is using the independent variable x.

Special Case 2. Uniform Load on the Entire Span

The purposes of presenting this very simple case are to show its relationships to the general case in the above, thereby enhancing the interconnectedness of problem-solving strategies through actual examples, as well as placing this important case in the proper context.

The reactions are

$$R_1 = \frac{wL}{2} = R_2 \tag{2.37}$$

The bending moments are

$$M = (R_1)x - \frac{wx^2}{2} \tag{2.38}$$

The maximum bending moment is

$$\max M = \frac{wL^2}{8} \quad \text{at } x = \frac{L}{2} \tag{2.39}$$

The deflections are

$$Y = \frac{1}{EI}\left[(S_1)x - \frac{1}{6}(R_1)x^3 + \frac{wx^4}{24}\right] \tag{2.40}$$

where

$$S_1 = \frac{wL^3}{24} \tag{2.41}$$

and R_1 is given by (2.37) above. Note that (2.40) is of exactly the same form as the corresponding formula in the general case presented earlier.

The maximum deflection is

$$\max Y = \frac{1}{EI}\left(\frac{5}{384}\right)wL^4 \quad \text{at } x = \frac{L}{2} \tag{2.42}$$

2.3. TRIANGULAR LOAD ON PART OF THE SPAN

General Case

The beam with its loading is shown in Figure 2.3.

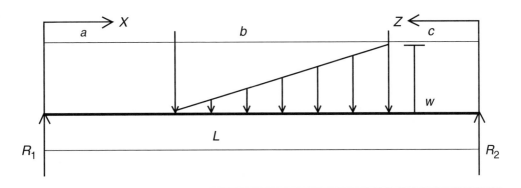

FIGURE 2.3. Triangular load on a simple beam

A triangular load can be represented by

$$p = q + rx \quad \text{with } q = \frac{-wa}{b}, \quad r = \frac{w}{b} \tag{2.43}$$

where w = maximum intensity of the triangular load.
The reactions are

$$R_1 = q\left[b - \frac{E_2}{L}\right] + r\left[E_2 - \frac{E_3}{L}\right] \tag{2.44}$$

$$R_2 = \left(\frac{1}{L}\right)[q(E_2) + r(E_3)] \tag{2.45}$$

with

$$E_2 = \frac{e^2 - a^2}{2}, \quad E_3 = \frac{e^3 - a^3}{3}, \quad e = a + b \tag{2.46}$$

The bending moments are

$$M_1 = (R_1)x \quad \text{for } 0 < x < a \tag{2.47}$$

$$M_2 = (R_1)x - B \quad \text{for } a < x < e \tag{2.48}$$

where

$$B = (x - a)^2\left[\frac{q}{2} + \left(\frac{r}{6}\right)(x + 2a)\right] \tag{2.49}$$

$$M_3 = (R_2)z \quad \text{for } 0 < z < c \tag{2.50}$$

The deflections are

$$Y_1 = \frac{1}{EI}\left[(S_1)x - \left(\frac{R_1}{6}\right)x^3\right] \quad \text{for } 0 < x < a \tag{2.51}$$

$$Y_2 = \frac{1}{EI}\left[(S_1)x - \left(\frac{R_1}{6}\right)x^3 + q\,\frac{(x-a)^4}{24} + r(x-a)^4\,\frac{x+4a}{120}\right] \tag{2.52}$$

for $a < x < e$

$$Y_3 = \frac{1}{EI}\left[(S_2)z - \left(\frac{R_2}{6}\right)z^3\right] \quad \text{for } 0 < z < c \tag{2.53}$$

Note that Formulae (2.51) and (2.53) are the same in form as the ones for the corresponding regions in the general case in Section 2.2, except that R_1 and R_2 are given by (2.44) and (2.45) respectively and S_1 and S_2 are

$$S_1 = C - \frac{D}{L} \tag{2.54}$$

$$S_2 = \frac{D}{L} \tag{2.55}$$

respectively, where

$$C = \frac{(R_1e^2 + R_2c^2)}{2} + \frac{wab^2}{6} - (wb^2)\,\frac{b+4a}{24} \tag{2.56}$$

$$D = (R_1)\,\frac{e^3}{3} + (R_2)c^2\,\frac{3L-2c}{6} + wab^2\,\frac{4a+3b}{24} \\ - (wb^2)\,\frac{5a^2 + 5ab + b^2}{30} \tag{2.57}$$

It is interesting to note also that Formula (2.52) in this section has some terms which are similar to those in Formula (2.23) in Section 2.2. In fact, the first three terms are exactly the same in form as before.

Special Case 1. a = 0

The reactions are

$$R_1 = wb\,\frac{3L-2b}{6L} \tag{2.58}$$

$$R_2 = \frac{wb^2}{3L} \tag{2.59}$$

The bending moments are

$$M_2 = (R_1)x - x^3\left[\frac{w}{6b}\right] \quad \text{for } 0 < x < b \tag{2.60}$$

$$M_3 = (R_2)z \quad \text{for } 0 < z < c \tag{2.61}$$

The maximum bending moments for the two regions are

$$\max M_2 = (R_1)(x_0) - (x_0)^3\left[\frac{w}{6b}\right] \quad \text{at } x_0 = \left[2b\frac{R_1}{w}\right]^{\frac{1}{2}} \tag{2.62}$$

$$\max M_3 = c(R_2) \quad \text{at } z = c \tag{2.63}$$

The deflections are

$$Y_2 = \frac{1}{EI}\left[(S_1)x - (R_1)\frac{x^3}{6} + \left(\frac{w}{b}\right)\frac{x^5}{120}\right] \quad \text{for } 0 < x < b \tag{2.64}$$

$$Y_3 = \frac{1}{EI}\left[(S_2)z - (R_2)\frac{z^3}{6}\right] \quad \text{for } 0 < z < c \tag{2.65}$$

with simplified

$$C = wb^2\frac{b + 4c}{24} \tag{2.66}$$

$$D = (R_1)\frac{b^3}{3} + (R_2)c^2\frac{3L - 2c}{6} - \frac{wb^4}{30} \tag{2.67}$$

The maximum deflections for the two regions are

$$\max Y_2 = \frac{1}{EI}h\left[S_1 - (R_1)\frac{h_2}{6} + \left(\frac{w}{b}\right)\frac{h^4}{120}\right] \tag{2.68}$$

$$\text{at } x = h \text{ for } 0 < x < b$$

where

$$h^2 = f - g, \quad f = 6\frac{R_1}{r}, \quad g = 2\frac{[9(R_1)^2 - 6r(S_1)]^{\frac{1}{2}}}{r} \tag{2.69}$$

$$\max Y_3 = \frac{1}{EI}\left[(S_2)(z_0) - (R_2)\frac{(z_0)^3}{6}\right] \quad \text{at } z_0 = \left[2\frac{S_2}{R_2}\right]^{\frac{1}{2}} \tag{2.70}$$

$$\text{for } 0 < z < c$$

Special Case 2. c = 0

The reactions are

$$R_1 = \frac{wb^2}{6L} \tag{2.71}$$

$$R_2 = w\frac{2L^2 - aL - a^2}{6L} \tag{2.72}$$

The bending moments are given by (2.47–2.49). Note that (2.50) is out in this case.

The deflections are given by (2.51) and (2.52) with $e = L$, and with R_1 and R_2 given by (2.71) and (2.45) respectively and

$$C = wb^2\frac{2L - b}{24} \tag{2.73}$$

$$D = wb^2\frac{20L^2 - 15ab - 12b^2}{360} \tag{2.74}$$

Note that the general Formulae (2.54) and (2.55) for S_1 and S_2 respectively still apply here, but with C and D given by (2.73) and (2.74) respectively.

2.4. TRIANGULAR LOAD ON THE ENTIRE SPAN

The reactions are

$$R_1 = \frac{wL}{6} \tag{2.75}$$

$$R_2 = \frac{wL}{3} \tag{2.76}$$

The bending moments are

$$M = (R_1)x - \left(\frac{w}{6L}\right)x^3 \tag{2.77}$$

The maximum bending moment is

$$\max M = \frac{wL^2}{9(3)^{1/2}} \quad \text{at } x = \frac{L}{(3)^{1/2}} \tag{2.78}$$

The deflections are

$$Y = \frac{1}{EI} \left(\frac{wx}{360L} \right) [3x^4 - 10(Lx)^2 + 7L^4] \tag{2.79}$$

where R_1 and R_2 are given by (2.75) and (2.76) respectively, and $S_1 = C - D/L$, $S_2 = D/L$ as in the general case, but with

$$C = \frac{wL^3}{24} \tag{2.80}$$

$$D = \frac{wL^4}{45} \tag{2.81}$$

2.5. GENERAL LOAD INTENSITY FUNCTIONS AND APPLICATIONS

Consider an arbitrary load intensity function $p(x)$. For most practical applications, it is sufficient to deal with elementary functions of a real variable (e.g., polynomials, trigonometric functions). In fact, in order to use the approach presented here for effective derivations of the formulae, $p(x)$ can be any function that can be integrated. The beam with a general load intensity function $p(x)$ is depicted in Figure 2.4.

Let

$$A = \int_a^e xp(x)dx, \qquad P = \int_a^e p(x)dx, \qquad e = a + b \tag{2.82}$$

The reactions are

$$R_1 = P - \frac{A}{L} \tag{2.83}$$

$$R_2 = \frac{A}{L} \tag{2.84}$$

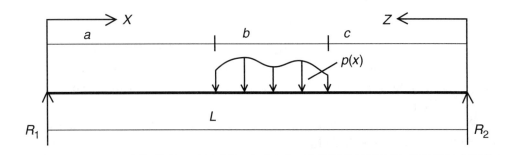

FIGURE 2.4. General load on a simple beam

The bending moments are

$$M_1 = (R_1)x \quad \text{for } 0 < x < a \tag{2.85}$$

$$M_2 = (R_1)x - B \quad \text{for } a < x < e \tag{2.86}$$

$$M_3 = (R_2)z \quad \text{for } 0 < z < c \tag{2.87}$$

where

$$B = \int_a^x (x - t)p(t)dt \tag{2.88}$$

Let

$$C = \int_0^a (R_1)x\,dx + \int_a^e [(R_1)x - B]dx + \int_0^c (R_2)z\,dz \tag{2.89}$$

$$D = \int_0^a (R_1)x^2\,dx + \int_a^e x[(R_1)x - B]dx + \int_0^c (L - x)(R_2)z\,dz \tag{2.90}$$

Then deflections are given by (2.91–2.93), with B, C, and D specified in (2.88– 2.90) respectively.

$$Y_1 = \frac{1}{EI}\left[(S_1)x - \left(\frac{1}{6}\right)(R_1)x^3\right] \quad \text{for } 0 < x < a \tag{2.91}$$

$$Y_2 = \frac{1}{EI}\left[(S_1)x - (R_1)a^2\,\frac{x - \frac{2a}{3}}{2} - \int_a^x (x - u)(M_2)(u)du\right] \tag{2.92}$$

for $a < x < e$

$$Y_3 = \frac{1}{EI}\left[(S_2)z - \left(\frac{1}{6}\right)(R_2)z^3\right] \quad \text{for } 0 < z < c \tag{2.93}$$

Note that (2.91) and (2.93) are exactly the same as the corresponding formulae for uniform and triangular load, over part of the span, which are special cases of the present section. Even Formula (2.92) bears some resemblance with the corresponding formulae for these load cases.

Note also that in the general case in Section 2.3 we see the same formal solution for deflections with the C, D functions for that particular load, namely triangular load. Similar remarks can be made about the uniformly distributed load over part of the span.

The general approach presented in this section is simple and efficient. The concept of general load intensity function can be and will be developed and applied to other beams including multi-span continuous beams to tackle classes of interesting and important problems.

2.6. A CONCENTRATED COUPLE AT AN ARBITRARY POINT ON THE SPAN

The beam with its loading is shown in Figure 2.5.

The reactions are

$$R_1 = -\frac{M_0}{L} \tag{2.94}$$

$$R_2 = \frac{M_0}{L} \tag{2.95}$$

The bending moments are

$$M = (R_1)x \quad \text{for } 0 < x < a \tag{2.96}$$

$$M = (R_2)z \quad \text{for } 0 < z < c \tag{2.97}$$

The maximum moment is

$$\max M = (R_1)a \quad \text{if } a > c \tag{2.98}$$

$$\max M = (R_2)c \quad \text{if } c > a \tag{2.99}$$

The deflections are

$$Y = \frac{1}{EI}\left[(M_0)\frac{x}{L}\right]\left(B + \frac{x^2}{6}\right) \quad \text{for } 0 < x < a \tag{2.100}$$

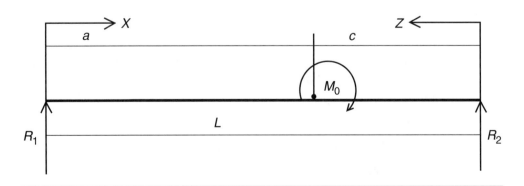

FIGURE 2.5. A concentrated couple at an arbitrary point on a simple beam

$$Y = \frac{1}{EI}\left[(M_0)\frac{z}{L}\right]\left(C + \frac{z^2}{6}\right) \quad \text{for } 0 < z < c \qquad (2.101)$$

where

$$B = \frac{1}{EI}\frac{1}{L}\left[\frac{c^3}{3} - \frac{a^2\left(c + \frac{a}{3}\right)}{2}\right] \qquad (2.102)$$

$$C = \frac{L^2 - a^2}{2} - B \qquad (2.103)$$

The deflection at the point of application of the concentrated couple is

$$Y = \frac{1}{EI}\left[a\,\frac{M_0}{L}\right]\frac{B - a^2}{6} \qquad (2.104)$$

or

$$Y = \frac{1}{EI}\left[c\,\frac{M_0}{L}\right]\frac{C - c^2}{6} \qquad (2.105)$$

2.7. THE PRINCIPLE OF SUPERPOSITION AND LOAD COMBINATIONS

Many important and useful load combinations are made available through the principle of superposition. The interested reader can formulate his or her own beam problems. This effective and simple approach will also be utilized for other types of structures including multi-span beams.

Example

Let us consider the beam shown in Figure 2.6. This is the load combination from the general cases in Sections 2.2 and 2.3. Here, the value of R_i in the resultant case is the algebraic sum of R_i in parts (1) and (2) shown in the figure for $i = 1, 2$.

2.8. EXPLORATIONS AND OBSERVATIONS

Section 2.1. The following remarks are offered regarding Section 2.1. The first interesting result for this simple problem is about deflections. We know that deflec-

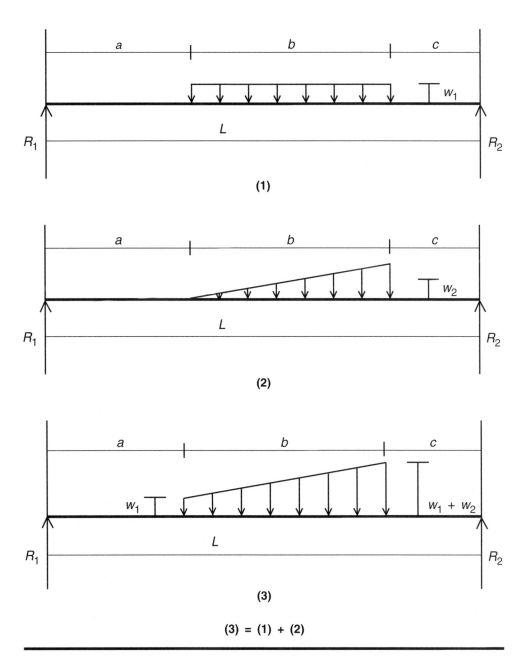

FIGURE 2.6. A simple beam example using the principle of superposition

tion is proportional to applied load naturally. Formula 2.4 shows that for the general case, deflection is also proportional to the maximum bending moment M_0 in the beam.

Next, the maximum deflection for the general case is proportional to, in addition to the factor $1/(EI)$, the following three factors: (1) the maximum bending moment, (2) a linear function of span length and location of applied load, as well as (3) the location of the maximum deflection. In this last statement, the first factor is very easy to see and comprehend. In fact, it is a natural extension of the general deflection mentioned in the last sentence of the previous paragraph. However, for the rest of the factors, it is not so obvious to see. Note also that these three factors listed above all have a common feature, namely that they are all of the dimension "length."

Moreover, again in the general case, the location of the maximum deflection does not coincide with the point of application of the concentrated force, as one might be inclined to think. It does so only when the single concentrated force is applied at mid-span. Another interesting thing to note is that the location of the maximum deflection depends only on span length and location of the applied load.

When a structure and applied load are given, the deflection, $X = A(xB)$, is a third-degree polynomial with $A = M_0/(EI)$, whereas B is a quadratic expression that consists of a constant and a second-degree term of the independent variable x. In other words, the quadratic expression does not contain a pure first-degree term of x. The amazing thing is that, as we will see later, this interesting phenomenon xB keeps on appearing in many other problems. This is true for the cases of a third-degree polynomial deflection function, a fourth-degree one, as well as a fifth-degree one. In fact, we will encounter one next.

Section 2.2. Several points are noted here. For the problem of a simple beam with a uniform load on one part of the span, deflection formulae for the loaded region and the rest of the span possess something in common. This amazing phenomenon is that they all have exactly the same third-degree polynomial characteristic mentioned in the general case, namely type xB. Of course, entity B here is the same for all regions in this section, and it is not identical to that in the general case. However, it is the same type that we are emphasizing. The difference between these regions for the same beam, as far as deflection is concerned, is that for the loaded region the deflection function is a fourth-degree polynomial in x, whereas for the rest of the span it is a third-degree one, the type of which is just as in Section 2.1. Note the similarity between the two sections in this regard. The additional term in the expression for deflection in the loaded region in this section signifies the "smoothing" effects of uniform load on the shape of the deflection curve.

From the above, we note that the third-degree polynomial characteristic xB mentioned above is significant in at least two aspects. First, it ties together loaded and unloaded regions of the beam in Section 2.2 in the manner just described. Second, it connects unloaded regions for both Sections 2.1 and 2.2 in the following way. Deflection functions for these regions all contain the fundamental ingredient, namely the xB-type polynomial. Furthermore, even for both the case of uniform load on the entire span and the case with only one unloaded region (for example, the case where $a = 0$), deflection functions for loaded regions are of the same form, namely a product xF of the independent variable x and a three-term third-degree polynomial F in x. Note that the interesting thing here is that F again does not contain a first-degree term of x either, just as mentioned before.

Section 2.3. For the problem of a simple beam with a triangular load on one part of the span, deflection functions for unloaded regions look exactly the same, in form, as in the previous two sections. This may be surprising to some people. As for loaded regions, a deflection function has one more term that represents a fifth degree in x, with the rest of the expression taking the same form as before in Section 2.2.

Consider the case of a load with the only unloaded region near the highest load intensity and the case of an entire span being loaded; the deflection function for the loaded region in each case is xG, the product of x and a fourth-degree polynomial G of the independent variable x. Interestingly enough, we see that G, following suit with B and F, does not contain a first-degree term of x either. By now, we see the interesting trend that deflection functions for different problems may have rather striking similarities after all.

Additionally, it is amazing to note that Formulae (2.51) and (2.53) themselves are of the same functional form when we review deflections in unloaded regions in the same problem.

Now let us return to the observation about bending moments for the case of a partially loaded span just cited. It is easy to see that the maximum bending moment in the loaded region as well as its location all depend on the reaction near the loaded region, the highest load intensity, and the length of the loaded region.

Section 2.4. This section deals with an important special case of a triangular load on the entire span. Concerning the deflections, the same remarks can be made as in Section 2.3 for a loaded region. However, the location of the maximum bending moment is a function of span length only. This is in sharp contrast to the slightly more general case, $a = 0$, where the location of the maximum bending moment depends on reactions, load characteristics, length of the loaded region, and other geometric features. Naturally, the contrast with the most general case is even more dramatic. Now, we get a feel for the effects of extent and location of load on the results of the solution to a problem.

Section 2.5. This section treats the general load intensity functions on the basis of the principle of superposition by using integration of results obtained from the case of a single concentrated force. The concept and approach as well as the notations remain essentially the same as in the problem presented in Section 2.1. Results for the special case with only one unloaded region can be obtained easily from the general case by letting the length of one of its unloaded regions be zero. This will simplify all formulae, especially those for deflections. For the case where the entire span is loaded, further simplification of results can be achieved and will be left to the reader to fill in the details.

Section 2.6. This section covers the topic of a simple beam with a concentrated couple at an arbitrary point. Some surprising things are noted. Deflection formulae have the form of the product of x^2 and a quadratic expression in x consisting of a constant term and x squared with a multiplier only. We note immediately that, as was seen before for an unloaded region of the beam, the polynomial has x as a factor. This universality property among various types of the so-called simple beams is very interesting.

Section 2.7. This section is about the principle of superposition, which facilitates finding many ways to obtain solutions to new problems from results of old problems. Some examples of different types of superposition are as follows.

One example is seen very often in practical applications. It is for a situation where the two original problems have the same given set of specified loaded regions. Thus, two different problems with their own loads are combined together naturally to form a new problem. The solution to the new problem is simply the algebraic sum of the solutions to the original problems.

There may be situations where loaded regions of one original problem are disjoint from those of the other original problem. Then the solution to the new problem to be obtained from superposition is simply the union of the solutions to the two original problems. There is, of course, superposition by integration as well as by combinations of types of superpositions mentioned above.

Naturally, there may be yet another load pattern in which the loaded region in one problem neither coincides with that in the other problem, nor are the loaded regions in the two given problems completely disjoint. Then, the way to deal with the situation is simply to break down the regions if needed and do our superposition task region by region so that we have either the first kind or the second kind in any one given region, where first kind or second kind means coincidence of the corresponding loaded regions in the two problems or being completely disjoint respectively.

3

CONTINUOUS BEAMS

3.1. TWO-SPAN CONTINUOUS BEAMS

3.1.1. A Concentrated Force at an Arbitrary Point on the Beam

Consider a two-span continuous beam with a concentrated force at any location, as shown in Figure 3.1. This is a statically indeterminate structure with one redundant reaction.

We will call this problem formally the "generic problem." The reason for it will be obvious as we develop the solution to this problem and see its implications and significance. Note also that we actually did make use of this idea in Chapter 2.

For convenience, we will take the exterior support reaction R of the loaded span as the unknown to be determined from the principle of strain energy or directly from

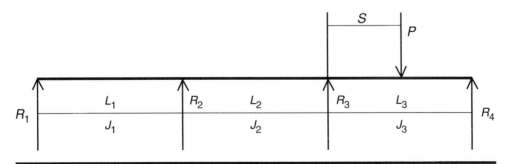

FIGURE 3.1. A concentrated force at an arbitrary point on a two-span continuous beam

the method of least work. Following the standard procedure in that method, we have the results as outlined below.

Noting that, from the solution to the problem, the reaction is proportional to the applied load P. We have

$$R = gP, \qquad R_1 = eP, \qquad R_2 = fP \tag{3.1}$$

where

$$e + f + g = 1, \qquad e = \frac{gL_2 - s}{L_1},$$

$$f = \frac{b - gL_0}{L_2}, \qquad L_0 = L_1 + L_2, \qquad b = L_1 + s \tag{3.2}$$

The bending moments for the three regions of the structure are

$$M_1 = ePx \quad \text{for } 0 < x < L_1 \tag{3.3}$$

$$M_2 = P[(gL_0 - b) + (1 - g)x] \quad \text{for } L_1 < x < b \tag{3.4}$$

and

$$M_3 = gP(L_0 - x) \quad \text{for } b < x < L_0 \tag{3.5}$$

We can rewrite R as

$$R = -\frac{AP}{B} \tag{3.6}$$

where

$$A = \left(\frac{s}{J_2}\right)\left[(L_0 + b)\frac{L_1 + b}{2} - (L_0 b + X) - (L_1)(L_2)\frac{J_2}{3J_1}\right] \tag{3.7}$$

$$B = \left(\frac{s}{J_2}\right)[L_0(L_2) - (L_0 b - X)] + (L_1)\frac{(L_2)^2}{3J_1} + \frac{c^3}{3J_2} \tag{3.8}$$

with

$$s + c = L_2, \qquad b = L_1 + s \tag{3.9}$$

$$X = \frac{b^2 + b(L_1) + (L_1)^2}{3} \tag{3.10}$$

Moreover, J_1 and J_2 are the product of Young's modulus and moment of inertia for spans 1 and 2 respectively.

Formulae (3.6–3.10) constitute the solution to the problem of finding the reaction R. With this and Formulae (3.1–3.5) we can obtain reactions R_1 and R_2 and the bending moments anywhere in the structure. Also, as usual, the determination of shear forces after bending moments are obtained is a simple matter of application of statics again.

Note that from Formulae (3.1) and (3.6) we have

$$g = -\frac{A}{B} \tag{3.11}$$

It turns out that using g instead of R is expedient as long as the load P remains constant.

Illustrative Example

Let's explore the solutions given above. The reason that we keep both formulae in (3.9) is simply for future reference when selecting the convenient (and also effective) parameters for further probing of the problem.

Note that both A and B are polynomials of the third degree in b, the location of the concentrated force with respect to the farthest exterior support for the given geometry. As long as the span lengths are fixed, it does not matter which one of the location designations b, c, and s shown in Figure 3.1 is selected as the parameter for further study. Observe also that A and B can be viewed as polynomials of the third degree in either c or s when L_1, L_2, J_1, and J_2 are given.

The next question is naturally something like this: What do Formulae (3.6–3.8) mean? To answer this question, first we will examine what the location parameter b can do to affect the reactions and, thus indirectly, the bending moments and shear forces. Applying the standard techniques of calculus for a continuous function of a real variable b, we note that we need to examine the function in (3.11).

As was mentioned above, A and B are both third-degree polynomials in b (or c or s). Which one of these three symbols to choose is entirely up to us. Observe that if we pick s as the variable to work with, then A is the product of s and a quadratic polynomial in s, while B will remain as a full-fledged third-degree polynomial with four terms. Let us do the analysis by utilizing the variable denoted by s as depicted in Figure 3.1. We have

$$A = s(hs^2 + js + k) \tag{3.12}$$

with coefficients h, j, and k looking very appealing. One of the reasons for taking this approach is that we are pretty comfortable with the product of the independent variable and a simple quadratic polynomial in the same variable as opposed to a general third-degree polynomial with four terms. So, let's decide to stick with this choice.

The next thing to do is to rewrite

$$B = ms^3 + ns^2 + ps + q \tag{3.13}$$

where m, n, p, and q are independent of s.

Thus, we have here a little project of investigating, mathematically, the ratio of two polynomials of the third degree with the numerator being of Formula (3.12). How do we go about doing this?

One way is to consider the function

$$F = \frac{A}{B} \tag{3.14}$$

and see under what conditions we will have an increasing or decreasing function or if we will have a maximum or minimum. If so, where? We can use our knowledge of calculus and set up the conditions for each of these considerations first and then write a mini-program and run it with results plotted in graphs if we so desire.

We deliberately define our function to be studied to look like what is given in (3.14), which just happens to be the negative of g given by (3.11). Therefore, we have to convert the results at the end by noting this. If we prefer, we could go ahead with the function g and name all coefficients of the polynomials accordingly to take care of the minus sign. No matter what choice we make, we are fine as long as we are consistent and careful in keeping track of all the details.

Now, can we see other features in the solution of R? We spot immediately the product of the moment of inertia and Young's modulus, the span lengths. Let's take care of J_1 and J_2 first. Let

$$r = \frac{J_1}{J_2} \tag{3.15}$$

for convenience, and we will choose at this time, for exploration, g as a function of r only. Thus

$$g = \frac{-A}{B} = \frac{ru + t}{rv + w} \tag{3.16}$$

where, t, u, v, and w are independent of J_1 and J_2 and thus also r. Thus, we are investigating g directly.

This is much easier than the calculation we did when the variable was s. We have the ratio between two linear algebraic functions. A purely analytic approach will do the job. We see that

$$\frac{dg}{dr} = (wu - tv)(rv + w)^{-2} \tag{3.17}$$

and it is positive, negative, or zero depending on whether

$$N = (wu - tv) \tag{3.18}$$

is positive, negative, or zero respectively.

Let us examine the packages t, u, v, and w and see what we have. We go back to Formulae (3.6), (3.7), and (3.15), and obtain

$$u = sL_1 \frac{L_2}{3} \tag{3.19}$$

$$v = (L_1) \frac{(L_2)^2}{3} \tag{3.20}$$

which are positive because L_1 and L_2 are, whereas the other two packages t and w are

$$t = \left[bL_0 + X - (b + L_0) \frac{b + L_1}{2} \right] s \tag{3.21}$$

$$w = [L_1 L_2 - (bL_0 - X)]s + \frac{c^3}{3} \tag{3.22}$$

Now let's see under what conditions N will be greater than, less than, or equal to zero.

We note that the expression N contains three parameters: L_1, L_2, and the location of the concentrated force. This location indicator can be represented by any one of the three symbols b, c, and s, depending on convenience in operation. Of course, they are interdependent as indicated in (3.9). Now, based on this set of information pieces, we can set up a program to determine the effects of r while holding L_1, L_2, and b constant. Note that we could use s instead of b, but do not have to.

We know when g will be an increasing function of r. Similarly, we can determine when g will be a decreasing function if there is such a case. Finally, what can we say about the possible extreme value of g? We look at $N = 0$ and its implications. This means that we solve the equation $N = 0$ for a specific parameter and check the results. We can do this easily by a simple program, which is nice. However, as it turns out, we can take an alternative approach by putting the solution to g into a different form, as illustrated in the next section.

3.1.2. An Alternative Approach to the Problem of Concentrated Force

Reference is made to Figure 3.1. Let us use as many new symbols as needed and start afresh. $R = gP$ is the same as before.

Recognizing the important lesson that we learned from using r above, let us be even more ambitious now. Let's define

$$m = \frac{L_1}{L_2} \tag{3.23}$$

$$n = \frac{J_1}{J_2} \tag{3.24}$$

for the reasons that we want to treat span #2 as the basis for comparison and we want to do things differently from what we did earlier.

Once we have our minds set, we do manipulations carefully by keeping packaging "airtight" and obtain the rather naive-looking result

$$g = \frac{ns(m + p)}{(m + n)L_2} \tag{3.25}$$

where

$$p = \frac{s(3L_2 - s)}{2L_2^2} \tag{3.26}$$

It is interesting to note that p is a functions of L_2 and s only. The exciting thing is that L_2 shows up in the definition of m and not as an independent entity anymore. This is a great advantage over the previous approach.

Let's rewrite g as

$$g = \frac{nf(s)}{(L_2)(m + n)} \tag{3.27}$$

where

$$f(s) = s(m + p) \tag{3.27A}$$

is regarded as a function of s only.

Something that is still more exciting is happening! g is in fact the product of two factors. One of them is $f(s)$, and the other is

$$Q = \frac{n}{(L_2)(m + n)} \tag{3.28}$$

which is clear of s. But what is Q? It is the ratio of T and L_2, where

$$T = \frac{n}{m + n} \tag{3.29}$$

Look at this new creature T. It is the ratio of n and $(m + n)$, and we have a very compact and meaningful package. This not only makes our work for computer applications much easier to do but also facilitates interpreting the results regarding physical significance.

On top of all that, we note that the function $f(s)$ is the product of s and a quadratic polynomial in s, because m is independent of s while p is itself the product of s and a linear function of s. In other words, $f(s)$ is a very special and simple third-degree polynomial. At this point, we can execute our program or pursue it a bit further analytically. The procedure for the latter is similar to what we did before, at least in principle and general directions. Either way, the total picture in front of us is very clear now.

For the sake of satisfying our curiosity, let us look at the special case specified by $m = 1$ and $n = 1$. We have, after some manipulations, the result

$$R = Ps \frac{4L^2 - c(L + c)}{4L^3} \tag{3.30}$$

where $c = L - s$, and $L_1 = L_2$ is called L here.

What happens if one of the distances a and c is zero or both are zero? Let us look at the results for several important special cases, as follows.

Special Case 1. $L_1 = L_2 = L$

We have simplifications immediately, with

$$m = 1 \tag{3.31}$$

and p is affected only by the fact that L_2 is replaced by L in (3.26). Therefore,

$$g = ns \frac{1 + p}{(1 + n)L} \tag{3.32}$$

Furthermore, if $s = 0$, then

$$g = 0 \tag{3.33}$$

If $s = L$, then

$$g = \frac{2n}{1 + n} \tag{3.34}$$

If $s = L/2$, then

$$g = \frac{\frac{13n}{16}}{1 + n} \tag{3.35}$$

Special Case 2. $J_1 = J_2 = J$

Here

$$n = 1 \tag{3.36}$$

p remains unchanged.

Hence,

$$g = s \frac{m + p}{(m + 1)L_2} \tag{3.37}$$

Furthermore, if $s = 0$, then

$$g = 0 \tag{3.38}$$

If $s = L_2$, then

$$g = 1 \tag{3.39}$$

If $s = L_2/2$, then

$$p = \frac{5}{8}, \qquad g = \frac{m + p}{2(m + 1)} \tag{3.40}$$

Special Case 3. $J_1 = J_2 = J$ and $L_1 = L_2 = L$

This means that both $m = 1$ and $n = 1$. Thus

$$g = s\frac{1 + p}{2L} \tag{3.41}$$

where p remains unchanged, as shown in (3.26).

Furthermore, if $s = 0$, then

$$g = 0 \tag{3.42}$$

If $s = L$, then

$$g = 1 \tag{3.43}$$

If $s = L/2$, then

$$p = \frac{5}{8} \tag{3.44}$$

$$g = \frac{13}{32} \tag{3.45}$$

3.1.3. Generic Problem, Arbitrary Load, and the Principle of Superposition: A Regression

Let us look at the principle of superposition by focusing on a list of some commonly used concentrated force cases:

1. Two or more concentrated forces on the same span
2. One concentrated force on each span
3. Two or more concentrated forces on each span

Let us extend the results just obtained for a concentrated force on a span to a few more cases of importance suggested above by the principle of superposition. In all cases, as far as the solution up to the level of all the reactions is concerned, this amounts to just simple algebraic addition of the solutions from the "parent" problems. "Algebraic addition" is mentioned because we may have upward vertical loads sometimes as well as the usual gravity loads when dealing with a horizontal beam. Moreover, we may want to study the beam behavior of a structural member which is not horizontally oriented and which is subjected to loads of a nature that may reverse direction.

What do we do about bending moments and shears? We have to be careful about the regions in which the formulae for moments, and hence shears, are valid. In other words, we have to watch out for the limits of validity of the formulae for bending moments and shear forces.

Speaking of the principle of superposition, we know that integration comes in handy when we are dealing with an arbitrary loading which is represented by a nice function. A "nice" function means that it is reasonably easy to handle. Thus, a nice function is necessarily continuous or at least piecewise continuous for our purposes here. Some examples are polynomials and sine and cosine functions, among many others. The functions represented by these examples can adequately represent the vast majority of the characteristics of the load intensity functions encountered in practical applications. What if for some reason we need to deal with discontinuous functions as our load intensity functions? We can break up the given function into several piecewise continuous functions. Each piece is continuous in a specific region. Then we can deal with the problem region by region, and in each region we have our nice function to work with, so there is no difficulty at all.

As to the terminology "generic problem," it simply means the following. It is a fundamental beam problem with its load type in the form of a single concentrated force at an arbitrary point on a span of a beam with one or more spans. We have seen a few examples before and will see more in the future, and that is why it is helpful to coin a term for it. The solution obtained from this generic problem forms the basis and starting point for solving the general or more complex problems with arbitrary load represented by a nice function as mentioned above. We will see an example below to elaborate further.

Illustrative Example: Arbitrarily Distributed Load and Revisiting the Problem in Section 3.1.2

By now, the reader may wonder how we proceeded to tackle the problem in Section 3.1.2 or, for that matter, having done several cases with arbitrarily distributed load since Chapter 2, may be curious about the answer to the same question just posed. Since our motivation is by now very strong, it appears that this is a good place and time to probe a little further along the same line. Let's start from the beginning.

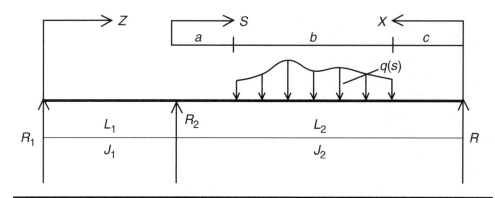

FIGURE 3.2. An arbitrarily distributed load on one span of a two-span continuous beam

Suppose we have an arbitrarily distributed load (see Figure 3.2). What that means is the following. The intensity, the extent, and the location of the loading are rather general but specified. In other words, we can deal with not only uniformly distributed load or linearly distributed load, but also many conceivably rather complicated load intensity functions.

Let's take a preview of the problem that we are going to solve to illustrate the point by going back to the core of the problem. We need to handle the integration of the function $f(s)q(s)$, where both $f(s)$ and $q(s)$ are considered as functions of s for our purpose here and now. As mentioned in Section 3.1.2, $f(s)$ is a simple third-degree polynomial in s. Now it is obvious that if we want to do the integration, we need $q(s)$ to be nice. Everything else in the expression g consists of factors independent of s and hence is merely a multiplier of the integration just indicated.

We now proceed to deal with the case of an arbitrary load on one span of a two-span continuous beam, given the load intensity function $q(s)$ of the arbitrary load. We look at the solution for a single concentrated force on a span (Section 3.1.2): $R = gP$. Now g as given by (3.27) in Section 3.1.2 is the product of T/L_2 and $f(s)$. Thus, for the present case, the integration to be carried over $gq(s)$ is in effect carried over $f(s)q(s)$ with a multiplier T/L_2 as forecast above.

We can see that the uniform load case is the case of $q(s) = w$, a constant. A triangular-shaped load intensity $q(s)$ is a linear function of s with a nonzero slope, with $q(s) = 0$ at a special point of the region of the load. Also, $q(s)$ equals a prescribed value at the end(s) of the region of the load. If someone can give us a nice $q(s)$, then we can provide an analytic solution in a few minutes. It is as simple as that once we have the solution to the "generic problem." But, of course, if we were to use a different form of the solution to the problem in Section 3.1.2, we might not be so lucky as to have an easy problem in this section. We realize the importance of doing mathematical operations efficiently. We also feel strongly that an effective innovative analytic treatment can really enhance our computer work substantially. To be sure, the amount of work is reduced and relevant parameters are revealed, with their significance easily assessed, clearly organized and displayed, ready for our use.

3.1.4. Uniform Load on One Span: General Case

The beam with its loading is shown in Figure 3.3. From the arbitrary load conditions, we can obtain the results for a uniform load over part of a span by setting $q(s) = w$, a constant, as follows

$$R = -\frac{L_1 L_2 \dfrac{A}{J_1} + \dfrac{3B}{J_2}}{D} \tag{3.46}$$

where

$$A = wb\left[c + \left(\frac{b}{2}\right) - L_2\right] \tag{3.47}$$

$$B = wb\left\{-\frac{L_2^3 - (b+c)^3}{3} + (2c+b)\frac{L_2^2 - (b+c)^2}{4} - (3b+4c)\frac{b^2}{24}\right\} \tag{3.48}$$

$$D = L_2^2\left(\frac{L_1}{J_1} + \frac{L_2}{J_2}\right) \tag{3.49}$$

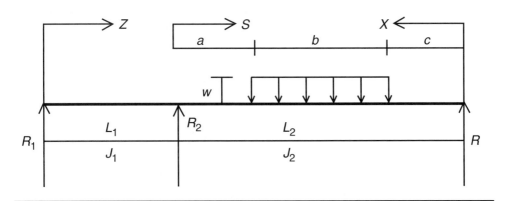

FIGURE 3.3. Uniform load on one span of a two-span continuous beam

The other reactions are obtained by the equations of equilibrium in statics as

$$R_1 = \frac{A + RL_2}{L_1} \tag{3.50}$$

$$R_2 = wb - \frac{A + FR}{L_1} \quad \text{with } F = L_1 + L_2 \tag{3.51}$$

Then the bending moments are obtained region by region in the usual manner.
For $0 < x < c$

$$M = Rx \tag{3.52}$$

For $c < x < (b + c)$

$$M = Rx - w\frac{(x - c)^2}{2} \tag{3.53}$$

For $b + c < x < L_2$

$$M = Rx - wb\left(x - c - \frac{b}{2}\right) \tag{3.54}$$

For $0 < z < L_1$

$$M = R_1 z \tag{3.55}$$

Special cases of importance under the general category of this section are covered
as follows.

Case 1. c = 0

This means that the load starts somewhere away from the interior support of the
span but extends to the exterior support of the loaded span. Therefore, $a + b = L_2$.
By setting $c = 0$ in the solution for the general uniform load problem, we have
immediately the simple results

$$A = wb\left[\frac{b}{2} - L_2\right] \tag{3.56}$$

$$B = wb\left[-\frac{L_2^3 - b^3}{3} + b\frac{L_2^2 - b^2}{4} - \frac{b^3}{8}\right] \tag{3.57}$$

D remains unchanged.

The other reactions are simplified from the corresponding expressions for the general uniform load over part of the span and are left to the reader to work out as exercises.

The bending moments are also simplified from the general uniform load problem as a direct result of setting $c = 0$.

Case 2. $a = 0$

This means that the loaded region starts from the interior support and ends at a point in the loaded span. Consequently, $b + c = L_2$.

Setting $a = 0$, we have the results

$$A = -\frac{wb^2}{2} \tag{3.58}$$

$$B = -(3L_2 + c)\frac{wb^3}{24} \tag{3.59}$$

D remains unchanged.

The bending moments are obtained by setting $a = 0$ in the corresponding expressions in the general uniform load case above.

Case 3. $a = c = 0$

This is the case where the entire span with length L_2 is uniformly loaded. The results can be obtained from combining cases 1 and 2 and noting $b = L_2$:

$$R = \frac{wL_2\left(\dfrac{4L_1}{J_1} + \dfrac{3L_2}{J_2}\right)}{8\left(\dfrac{L_1}{J_1} + \dfrac{L_2}{J_2}\right)} \tag{3.60}$$

R_1 and R_2 are given by (3.50) and (3.51) respectively.

The bending moments are also further simplified compared with cases 1 and 2 above. For $0 < x < L_2$

$$M = Rx - \frac{wx^2}{2} \tag{3.61}$$

For $0 < z < L_1$

$$M = R_1 z \tag{3.62}$$

3.1.5. Uniform Load on One Span: Special Case with Equal Span Lengths

Reference is made to Figure 3.3, with the special provision that $L_1 = L_2 = L$. In other words, the material and section properties are still quite general in that J_1 and J_2 are different.

The results are

$$R = \frac{-\dfrac{L^2 A}{J_1} + \dfrac{3B}{J_2}}{D} \tag{3.63}$$

where

$$A = wb\left(c + \frac{b}{2} - L\right) \tag{3.64}$$

$$B = \left\{ \begin{array}{l} -\dfrac{L^3 - (b + c)^3}{3} + (2c + b)\dfrac{L^2 - (b + c)^2}{4} \\[2mm] -(3b + 4c)\dfrac{b^2}{24} \end{array} \right\} wb \tag{3.65}$$

$$D = L^3\left(\frac{1}{J_1} + \frac{1}{J_2}\right) \tag{3.66}$$

$$R_1 = \frac{A + RL}{L} \tag{3.67}$$

$$R_2 = wb - \frac{A + 2LR}{L} \tag{3.68}$$

The bending moments remain unchanged from those in Section 3.1.4, but with R and R_1 given by results from this section.

Special cases of importance under the general category of this section are covered as follows.

Case 1. c = 0

$$B = wb\left[-\frac{L^3 - b}{3} + b\frac{L^2 - b^2}{4} - \frac{b^3}{8}\right] \tag{3.69}$$

A and D remain unchanged from case 1 in Section 3.1.4.

$$R = -\frac{\dfrac{L^2 A}{J_1} + \dfrac{3B}{J_2}}{D} \tag{3.70}$$

R_1 and R_2 are given by (3.67) and (3.68) respectively, with R given by (3.70).

Case 2. a = 0

$$A = -\frac{wb^2}{2} \tag{3.71}$$

$$B = -(3L + c)\frac{wb^3}{24} \tag{3.72}$$

D remains unchanged. R, R_1, and R_2 are given by (3.63), (3.67), and (3.68) respectively.

Case 3. a = c = 0

$$R = wL\,\frac{\dfrac{4}{J_1} + \dfrac{3}{J_2}}{8\left(\dfrac{1}{J_1} + \dfrac{1}{J_2}\right)} \tag{3.73}$$

R_1 and R_2 are given by (3.67) and (3.68) respectively where

$$A = -\frac{wL^2}{2}, \qquad B = -\frac{wL^4}{8} \tag{3.74}$$

The bending moments are also further simplified compared with cases 1 and 2 above. For $0 < x < L$

$$M = Rx - \frac{wx^2}{2} \tag{3.75}$$

For $0 < z < L$

$$M = R_1 z \tag{3.76}$$

3.1.6. Uniform Load on One Span: Special Case with Constant Material and Section Properties

Reference is made to Figure 3.3 again. This time, the special provision is $J_1 = J_2 = J$. Note that here no restrictions are imposed on the magnitude of L_1 and L_2. Therefore, the results are the same as the general case in Section 3.1.4, except that J_1 and J_2 are to be replaced by J.

The interested reader can, out of curiosity, write out the explicit expressions of the resulting solutions by following the guidelines mentioned above. To facilitate checking the results of self-study, an example is provided below. This simple but important example can also serve the additional purpose of arousing interest on the part of the reader to actively engage in probing with optimal efficiency. Thus, this brief section provides an opportunity to do so at a suitable point in our development of topics.

For the special case where $a = 0 = c$, and thus $b = L_2$, we have

$$R = wL_2 \frac{4L_1 + 3L_2}{8(L_1 + L_2)} \tag{3.77}$$

R_1 and R_2 are given formally by (3.50) and (3.51) respectively in Section 3.1.4 with simplified expressions

$$A = -\frac{wb^2}{2}, \quad B = -\frac{wb^4}{8} \tag{3.78}$$

The formulae for bending moments remain the same as in Section 3.1.4.

3.1.7. Uniform Load on One Span: Special Case with Constant Material and Geometric Characteristics

The results are

$$R = -\frac{L^2 A + 3B}{JD} \tag{3.79}$$

where

$$A = wb\left[c + \frac{b}{2} - L\right] \tag{3.80}$$

$$B = \left\{ \begin{array}{l} -\dfrac{L^3 - (b + c)^3}{3} + (2c + b)\dfrac{L^2 - (b + c)^2}{4} \\ -(3b + 4c)\dfrac{b^2}{24} \end{array} \right\} wb \tag{3.81}$$

$$D = \frac{2L^3}{J} \tag{3.82}$$

The other reactions are given formally in Section 3.1.4, and the bending moments are the same as in Section 3.1.4.

Special cases of importance under the general category of this section are covered as follows.

Case 1. c = 0

$$A = wb\left(\frac{b}{2} - L\right) \tag{3.83}$$

$$B = wb\left[-\frac{L^3 - b^3}{3} + b\frac{L^2 - b^2}{4} - \frac{b^3}{8}\right] \tag{3.84}$$

D remains unchanged as shown in (3.82). R is given by (3.79). R_1 and R_2 are given formally by (3.50) and (3.51) respectively in Section 3.1.4.

Case 2. a = 0

$$A = -\frac{wb^2}{2} \tag{3.85}$$

$$B = -(3L + c)\frac{wb^3}{24} \tag{3.86}$$

D remains unchanged. R is given by (3.79). R_1 and R_2 are given formally by (3.50) and (3.51) respectively in Section 3.1.4.

Case 3. a = c = 0

$$R = \frac{7wL}{16} \tag{3.87}$$

$$R_1 = \frac{-wL}{16} \tag{3.88}$$

$$R_2 = \frac{5wL}{8} \tag{3.89}$$

The bending moments are the same formally as in Section 3.1.4.

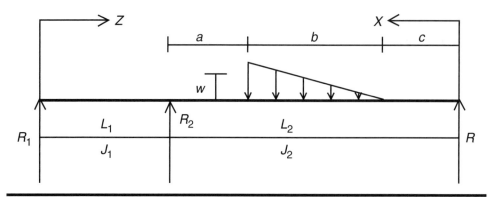

FIGURE 3.4. Triangular load on one span of a two-span continuous beam

3.1.8. Triangular Load on One Span: General Case

The beam with its loading is shown in Figure 3.4. From the arbitrary load conditions, we can obtain the results for a triangular load over part of a span by setting $q(s)$ to be a linear function of s, or we can derive the formulae for the solution directly.

The results are

$$R = -\frac{\dfrac{L_1 L_2 A}{J_1} + \dfrac{3B}{J_2}}{D} \tag{3.90}$$

where

$$A = wb\,\frac{c + \dfrac{2b}{3} - L_2}{2} \tag{3.91}$$

$$B = \left\{ \begin{array}{l} (2b + 3c)\,\dfrac{L_2^2 - (b + c)^2}{12} - \dfrac{L_2^3 - (b + c)^3}{6} \\[2mm] -\,(4b - 5c)\,\dfrac{b^2}{120} \end{array} \right\} wb \tag{3.92}$$

$$D = L_2^2\left(\frac{L_1}{J_1} + \frac{L_2}{J_2}\right) \tag{3.93}$$

Note that both A and B contain only the entities w, b, c, and L_2, while the expression D contains L_2 as well as the relative flexibilities L_1/J_1 and L_2/J_2 for the two spans.

The other reactions are obtained by the equations of equilibrium in statics and are

$$R_1 = \frac{A + RL_2}{L_1} \tag{3.94}$$

$$R_2 = \frac{wb}{2} - \frac{A + RF}{L_1} \quad \text{with } F = L_1 + L_2 \tag{3.95}$$

Then, the bending moments are obtained region by region in the usual manner as follows. For $0 < x < c$

$$M = Rx \tag{3.96}$$

For $c < x < (b + c)$

$$M = Rx - w\frac{(x - c)^3}{6b} \tag{3.97}$$

For $b + c < x < L_2$

$$M = Rx - wb\frac{x - \left(c + \dfrac{2b}{3}\right)}{2} \tag{3.98}$$

For $0 < z < L_1$

$$M = R_1 z \tag{3.99}$$

Special cases of importance under the general category of this section are covered below.

Case 1. c = 0

This means that the load starts somewhere away from the interior support of the span and extends to the exterior support of the loaded span. Therefore, $a + b = L_2$.

By setting $c = 0$ in the solution for the general uniform load problem, we have immediately the simple results

$$A = \frac{wb\left(\dfrac{2b}{3} - L_2\right)}{2} \tag{3.100}$$

$$B = \left[b\frac{L_2^2 - b^2}{6} - \frac{L_2^3 - b^3}{6} - \frac{b^3}{30}\right]wb \tag{3.101}$$

R is given by (3.90). D remains unchanged. The other reactions are simplified from the corresponding expressions above for the general triangular load over part of the span and are left to the reader to work out as exercises.

The bending moments are also simplified from the general triangular load problem as a direct result of setting $c = 0$.

Case 2. *a* = 0

This means that the loaded region starts from the interior support and ends at a point in the loaded span. Consequently, $b + c = L_2$.

Setting $a = 0$, we have the results

$$A = \frac{-wb^2}{6} \tag{3.102}$$

$$B = -(4b - 5c)\frac{wb^3}{120} \tag{3.103}$$

D remains unchanged. R, R_1, and R_2 are given by (3.90), (3.94), and (3.95) respectively.

The bending moments are obtained by setting $a = 0$ in the corresponding expressions in the general triangular load case above.

Case 3. *a* = *c* = 0

This is the case where the entire span with length L_2 is under triangular load. The results can be obtained from combining cases 1 and 2 and noting $b = L_2$:

$$R = \frac{wL_2\left(\dfrac{5L_1}{J_1} + \dfrac{3L_2}{J_2}\right)}{30\left(\dfrac{L_1}{J_1} + \dfrac{L_2}{J_2}\right)} \tag{3.104}$$

R_1 and R_2 are given by (3.94) and (3.95) respectively with

$$A = -\frac{wL_2^2}{6}, \qquad B = -\frac{wL_2^4}{30} \tag{3.105}$$

The bending moments are as follows. For $0 < x < L_2$

$$M = Rx - \frac{wx^3}{6L_2} \tag{3.106}$$

For $0 < z < L_1$

$$M = R_1 z \tag{3.107}$$

3.1.9. Triangular Load on One Span: Special Case with Equal Span Lengths

Reference is made to Figure 3.4 with the special provision of $L_1 = L_2 = L$.

$$R = - \frac{\dfrac{L^2 A}{J_1} + \dfrac{3B}{J_2}}{D} \tag{3.108}$$

where

$$A = wb \frac{c + \dfrac{2b}{3} - L}{2} \tag{3.109}$$

$$B = \left\{ \begin{array}{l} (2b + 3c) \dfrac{L^2 - (b + c)^2}{12} - \dfrac{L^3 - (b + c)^3}{6} \\[2ex] - (4b - 5c) \dfrac{b^2}{120} \end{array} \right\} wb \tag{3.110}$$

$$D = L^3 \left(\frac{1}{J_1} + \frac{1}{J_2} \right) \tag{3.111}$$

The other reactions are obtained by the equations of equilibrium in statics and are

$$R_1 = \frac{A + RL}{L} \tag{3.112}$$

$$R_2 = \frac{wb}{2} - \frac{A + 2RL}{L} \tag{3.113}$$

The bending moments are given by (3.96–3.99) in Section 3.1.8, with R and R_1 given by (3.108) and (3.112) respectively.

Special cases of importance under the general category of this section are covered as follows.

Case 1. c = 0

$$A = \frac{wb \left(2 \dfrac{b}{3} - L \right)}{2} \tag{3.114}$$

$$B = \left[b \frac{L^2 - b^2}{6} - \frac{L^3 - b^3}{6} - \frac{b^3}{30} \right] wb \tag{3.115}$$

D remains unchanged and is shown in (3.111). R is formally as shown in (3.108), with A and B given by (3.114) and (3.115) respectively.

Case 2. a = 0

$$A = -\frac{wb^2}{6} \tag{3.116}$$

$$B = -(4b - 5c)\frac{wb^3}{120} \tag{3.117}$$

D remains unchanged. R, R_1, and R_2 are given formally in (3.108), (3.112), and (3.113) respectively, with A and B given in (3.116) and (3.117) respectively.

Case 3. a = c = 0

This is the case where the entire span #2 (length L) is under triangular load. The results can be obtained from combining cases 1 and 2 and noting $b = L$.

$$R = wL\frac{\dfrac{5}{J_1} + \dfrac{3}{J_2}}{30\left(\dfrac{1}{J_1} + \dfrac{1}{J_2}\right)} \tag{3.118}$$

R_1 and R_2 are given formally by (3.112) and (3.113) respectively, with A and B given by

$$A = -\frac{wL^2}{6}, \qquad B = -\frac{wL^4}{30} \tag{3.119}$$

The bending moments are as follows. For $0 < x < L$

$$M = Rx - \frac{wx^3}{6L} \tag{3.120}$$

For $0 < z < L$

$$M = R_1 z \tag{3.121}$$

3.1.10. Triangular Load on One Span: Special Case with Constant Material and Section Properties

Reference is made to Figure 3.4, with the special provision of $J_1 = J_2 = J$. The results are

$$R = -\frac{L_1 L_2 A + 3B}{JD} \tag{3.122}$$

with A and B given by (3.91) and (3.92) respectively in Section 3.1.8.

$$D = L_2^2 \left(\frac{L_1}{J} + \frac{L_2}{J} \right) \tag{3.123}$$

Note that both A and B contain only the entities w, b, c, and L_2, while the expression D contains L_2 as well as the relative flexibilities L_1/J and L_2/J for the two spans.

The reactions R_1 and R_2 are given by (3.94) and (3.95) respectively in Section 3.1.8, with R given by (3.122).

The bending moments are given by expressions in Section 3.1.8, with R and R_1 given by expressions in this section.

Special cases of importance under the general category of this section are covered as follows.

Case 1. c = 0

The results are obtained from case 1 in Section 3.1.8 by letting $J_1 = J_2 = J$.

Case 2. a = 0

The results are the same, formally, as in case 2 in Section 3.1.8 except that R and R_1 are given by expressions in this section.

Case 3. a = c = 0

$$R = wL_2 \, \frac{5L_1 + 3L_2}{30(L_1 + L_2)} \tag{3.124}$$

R_1 and R_2 are given formally by (3.94) and (3.95) respectively, with A and B given by (3.105) in Section 3.1.8.

The bending moments are given formally by Formulae (3.106) and (3.107) in Section 3.1.8.

3.1.11. Triangular Load on One Span: Special Case with Constant Material and Geometric Characteristics

Reference is made to Figure 3.4 with the special provision of $J_1 = J_2 = J$ and $L_1 = L_2 = L$.

The results are

$$R = -\frac{L^2 A + 3B}{JD} \qquad (3.125)$$

where

$$A = wb \frac{c + \dfrac{2b}{3} - L}{2} \qquad (3.126)$$

$$B = \left\{ \begin{array}{l} (2b + 3c)\dfrac{L^2 - (b + c)^2}{12} - \dfrac{L^3 - (b + c)^3}{6} \\[2ex] - (4b - 5c)\dfrac{b^2}{120} \end{array} \right\} wb \qquad (3.127)$$

$$D = \frac{2L^3}{J} \qquad (3.128)$$

Note that both A and B contain only the entities w, b, c, and L, while the expression D contains L as well as the relative flexibility L/J for the two spans.

The other reactions are obtained by the equations of equilibrium in statics and are

$$R_1 = \frac{A + RL}{L} \qquad (3.129)$$

$$R_2 = \frac{wb}{2} - \frac{A + 2RL}{L} \qquad (3.130)$$

Then, the bending moments are obtained region by region in the usual manner as follows. For $0 < x < c$

$$M = Rx \qquad (3.131)$$

For $c < x < (b + c)$

$$M = Rx - w\frac{(x - c)^3}{6b} \qquad (3.132)$$

For $b + c < x < L$

$$M = Rx - wb \frac{x - \left(c + \dfrac{2b}{3}\right)}{2} \qquad (3.133)$$

For $0 < z < L$

$$M = R_1 z \qquad (3.134)$$

Special cases of importance under the general category of this section are covered as follows.

Case 1. c = 0

This means that the load starts somewhere away from the interior support on the span and extends to the exterior support of the loaded span. Therefore, $a + b = L$.

By setting $c = 0$ in the solution for the general problem, we have immediately the simple results

$$A = \frac{wb\left(\dfrac{2b}{3} - L\right)}{2} \tag{3.135}$$

$$B = \left[b \frac{L^2 - b^2}{6} - \frac{L^3 - b^3}{6} - \frac{b^3}{30}\right]wb \tag{3.136}$$

D remains unchanged and is shown in (3.128). R is formally as shown in (3.125), with A and B given by (3.135) and (3.136) respectively.

The other reactions are simplified from the corresponding expressions above for the general problem with varying span lengths as well as material and section properties and are left to the reader to work out as exercises.

The bending moments are also simplified from the general problem as a direct result of setting $c = 0$ in the expressions for R and R_1 which enter into the expressions for bending moments.

Case 2. a = 0

This means that the loaded region starts from the interior support and ends at an interior point on the loaded span. Consequently, $b + c = L$.

Setting $a = 0$, we have the results

$$A = -\frac{wb^2}{6} \tag{3.137}$$

$$B = -(4b - 5c)\frac{wb^3}{120} \tag{3.138}$$

D remains unchanged. R, R_1, and R_2 are given formally in (3.125), (3.129), and (3.130) respectively, with A and B given in (3.137) and (3.138) respectively.

The bending moments are obtained by using R and R_1 just obtained in the corresponding expressions for bending moments in the general case above in this section.

Case 3. a = c = 0

This is the case where the entire span #2 (length L) is under triangular load. The results can be obtained from combining cases 1 and 2 and noting $b = L$

$$R = \frac{2wL}{15} \tag{3.139}$$

R_1 and R_2 are given formally by (3.129) and (3.130) respectively, with A and B given by

$$A = -\frac{wL^2}{6}, \qquad B = -\frac{wL^4}{30} \tag{3.140}$$

The bending moments are also further simplified compared with cases 1 and 2 above. For $0 < x < L$

$$M = Rx - \frac{wx^3}{6L} \tag{3.141}$$

For $0 < z < L$

$$M = R_1 z \tag{3.142}$$

3.1.12. A Concentrated Couple at an Arbitrary Point on a Span

As shown in Figure 3.5, the exterior end reaction R for the loaded span is taken as the unknown in the equation resulting from the method of least work. The solution is given by

$$R = \frac{AM_0}{D} \tag{3.143}$$

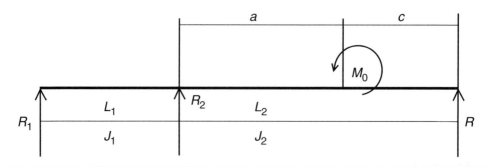

FIGURE 3.5. A concentrated couple at an arbitrary point on a two-span continuous beam

where

$$A = 3 \left[\frac{C_1 L_2}{3} + \frac{L_2^2 - c^2}{2 J_2} \right] \tag{3.144}$$

$$D = L_2^2 (C_1 + C_2) \tag{3.145}$$

with

$$C_i = \frac{L_i}{J_i}, \quad i = 1, 2 \tag{3.146}$$

Note that the definitions for D and C_i are as before in this chapter, as they should be.

It is apparent from (3.143) and (3.144) that the effects of c on R can be seen by looking at the function

$$R = k - h c^2 \tag{3.147}$$

where k and h are independent of c.

It is easy to see that when $c = 0$, R has one extreme value of

$$S_1 = 3 \left(\frac{C_1 L_2}{3} + \frac{C_2 L_2}{2} \right) \frac{M_0}{D} \tag{3.148}$$

Also, when $c = L_2$, the reaction R has the other extreme value of

$$S_2 = \frac{-C_1 L_2 M_0}{D} \tag{3.149}$$

We observe from (3.143) that R is directly proportional to the applied load, namely the concentrated couple.

As for the effects of C_1, C_2 and those of L_2 and J_2 separately, we can run simple programs using a spreadsheet approach to obtain meaningful results. The interested reader can use Formula (3.143) as a start and get involved right away.

Once again, the other reactions as well as the bending moments and shear forces can be obtained easily from the equations of statics and are therefore left to the interested reader to pursue.

3.2. THREE-SPAN CONTINUOUS BEAMS

3.2.1. A Concentrated Force at an Arbitrary Point on an Exterior Span

The beam with its loading is shown in Figure 3.6. Just as with two-span continuous beams, this load case is important for two reasons. First, it is an adequate and useful

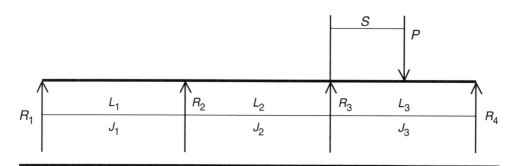

FIGURE 3.6. A concentrated force at an arbitrary point on an exterior span of a three-span continuous beam

representation of an actual situation. Second, its solution serves the additional purpose of constructing solutions for a large class of important load cases via the principle of superposition in its broadest sense, which includes the method of integration for obtaining solutions for arbitrarily distributed loads.

By selecting reactions R_1 and R_4 as the unknowns in the system of simultaneous equations resulting from the method of least work, we have the following solutions

$$R_1 = Z(sT + KL_3)\frac{P}{D} \tag{3.150}$$

$$R_4 = -(Z^2 sL_3 + KG)\frac{P}{D} \tag{3.151}$$

where

$$Z = \frac{L_1 L_2}{6J_2} \tag{3.152}$$

$$T = (C_2 + C_3)\frac{L_3^2}{3} \tag{3.153}$$

$$K = s\frac{\dfrac{s^2}{J_3} - 3sC_3 - 2C_2 L_3}{6} \tag{3.154}$$

$$G = (C_1 + C_2)\frac{L_1^2}{3} \tag{3.155}$$

$$C_i = \frac{L_i}{J_i}, \quad i = 1, 2, 3 \tag{3.156}$$

$$D = \left[\frac{L_1 L_3}{6}\right]^2 [3C_2^2 + 4(C_1 + C_3)C_2 + 4C_1 C_3] \qquad (3.157)$$

The denominator D in R_1 and R_4 characterizes the structure in terms of material and section properties and span lengths, or in plain language, it is a measure of the capability of the structure, and it will appear again in other three-span continuous beam problems. Recall that in Section 3.1, we saw a similar thing, also called D there, which is simpler in form than the present one.

For convenience in future applications, we may rewrite R_1 and R_4 as

$$R_1 = (As + BK)\frac{P}{D} \qquad (3.158)$$

$$R_4 = (Us - GK)\frac{P}{D} \qquad (3.159)$$

where A, B, D, G, and U are independent of s, with

$$A = ZT, \qquad B = ZL_3, \qquad U = -Z^2 L_3 \qquad (3.160)$$

Note that G and D are defined in (3.155) and (3.157) respectively, while K, as defined in (3.154), is a third-degree polynomial and is a product of s and a quadratic polynomial of s.

Let us look at some interesting and important special cases.

Case 1. Equal Span Lengths

Using $L_1 = L_2 = L_3 = L$ in the solutions for the general case, we obtain R_1 and R_4 as shown in (3.150) and (3.151) respectively, with

$$Z = \frac{L^2}{6J_2} \qquad (3.161)$$

$$T = (C_2 + C_3)\frac{L^2}{3} \qquad (3.162)$$

$$K = s\,\frac{\dfrac{s^2}{J_3} - 3sC_3 - 2C_2 L}{6} \qquad (3.163)$$

$$G = (C_1 + C_2)\frac{L^2}{3} \qquad (3.164)$$

$$C_i = \frac{L_i}{J_i}, \quad i = 1, 2, 3 \qquad (3.165)$$

$$D = L^4 \frac{3C_2^2 + 4(C_1 + C_3)C_2 + 4C_1C_3}{36} \tag{3.166}$$

The other reactions are simplified accordingly also.

Case 2. $J_1 = J_2 = J_3 = J$

$$Z = \frac{L_1 L_2}{6J} \tag{3.167}$$

$$T = (C_2 + C_3)\frac{L_3^2}{3} \tag{3.168}$$

$$K = s\frac{\dfrac{s^2}{J} - 3sC_3 - 2C_2L_3}{6} \tag{3.169}$$

Formulae (3.155) and (3.157) in the general case hold formally here. However, Formula (3.156) becomes

$$C_i = \frac{L_i}{J}, \quad i = 1, 2, 3 \tag{3.170}$$

The reactions R_1 and R_4 are as shown in (3.150) and (3.151) respectively, with components given in (3.167–3.170) above.

Case 3. $L_1 = L_2 = L_3 = L$ and $J_1 = J_2 = J_3 = J$
We have

$$Z = \frac{L^2}{6J} \tag{3.171}$$

$$T = \frac{2L^2}{3J} \tag{3.172}$$

$$K = s\frac{\dfrac{s^2}{J} - 3Cs - 2CL}{6} \tag{3.173}$$

$$G = \frac{2L^3}{3J} = \frac{2CL^2}{3} \tag{3.174}$$

$$C = \frac{L}{J} \tag{3.175}$$

$$D = 15L^4 \, \frac{\left(\dfrac{L}{J}\right)^2}{36} = \frac{15L^4 C^2}{36} \tag{3.176}$$

Illustrative Example

So far in this section, we have seen two different forms of solution to the same problem of a single concentrated force on one span. We call this problem the generic problem also because of its capability by which we can derive or construct solutions to other problems systematically. Each solution form has its particular features and thus particular charms and power. Once several forms are available, we can choose one of them for future use, depending on what we want to achieve.

Formulae (3.158) and (3.159) are of the form

$$R = \frac{NP}{D} \tag{3.177}$$

where

$$N = As + BK \quad \text{for } R_1 \tag{3.178}$$

$$N = Us - GK \quad \text{for } R_4 \tag{3.179}$$

and D is given by (3.157).

We note that D contains two factors. One is $(L_1 L_3/6)^2$, and the other is the sum of quadratic terms of C_i, $i = 1, 2, 3$.

The solution to a two- or three-span continuous beam problem has basically two parts. One part F specifies the load-related information, and the other part H has to do with material and section properties as well as other geometric characteristics. Hence, for the theme of this chapter, we have

$$R = FH \tag{3.180}$$

Here H itself takes the form of a ratio of two entities, as will be seen below.

Comparing (3.177) with (3.180), we see that

$$PN = FQ, \qquad \frac{Q}{D} = H \tag{3.181}$$

Thus,

$$R = \frac{FQ}{D} \tag{3.182}$$

Of course, we can see directly from the onset that $R = FQ/D$ also.

Which one of these approaches we should take is a matter of personal preference and convenience. We are liberal about our choice even in mathematical matters. Naturally, it has to be within reason. After all, mathematics entails adherence to and application of logical reasoning. There is freedom of choice, but after the choice is made, we must stick to it all the way for consistency and correctness.

Note that F depends on the magnitude and location of load. What we do to derive the solution to a problem with any type of distributed load is to treat F from the generic problem as we have already seen in a few examples.

Let's use the form of solution in (3.182). Then

$$R_1 = \frac{F_1 Q_1}{D} \tag{3.183}$$

$$R_4 = \frac{F_4 Q_4}{D} \tag{3.184}$$

where

$$F_1 = P\left[s + \left(\frac{B}{A}\right)K\right] \quad \text{with} \quad \frac{B}{A} = \frac{3L_3}{L_2^3} \tag{3.185}$$

$$F_4 = P\left[s - \left(\frac{G}{U}\right)K\right] \tag{3.186}$$

$$Q_1 = A \tag{3.187}$$

$$Q_4 = U \tag{3.188}$$

with A, B, G, K, and U defined earlier.

F_1 and F_4 are products of s and a quadratic polynomial in s, as we observed earlier using different notations.

3.2.2. Arbitrarily Distributed Load on an Exterior Span

The beam with its loading is shown in Figure 3.7. For convenience and consistency, let's continue to use the notations

$$R_1 = \frac{F_1 Q_1}{D} \tag{3.189}$$

$$R_4 = \frac{F_4 Q_4}{D} \tag{3.190}$$

where F_1 and F_4 depend on load, while Q_1, Q_4, and D are independent of load.

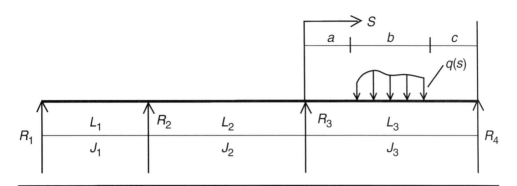

FIGURE 3.7. An arbitrarily distributed load on an exterior span of a three-span continuous beam

F_1 and F_4 are obtained from integrating the effects of load intensity function $q(s)$ over the loaded region by the method illustrated earlier. Note that, in the process of integration indicated above, Q_1/D and Q_4/D are simple factors of multiplication. The specific expressions for component factors of the solutions are

$$F_1 = \int_a^e f_1(s)q(s)ds \tag{3.191}$$

$$F_4 = \int_a^e f_4(s)q(s)ds \tag{3.192}$$

where

$$f_1(s) = s + \frac{KB}{A} \tag{3.193}$$

$$f_4(s) = s - \frac{KG}{U} \tag{3.194}$$

with A, B, G, K, U, Q_1, Q_4, and D defined in Section 3.2.1.

Let us look at some interesting and important special cases in the class of arbitrary load intensity functions. The formal solutions are still given by Expressions (3.189) and (3.190), with Q_1 and Q_4 given earlier in this chapter. However, the expressions for F_1 and F_4 are simplified as follows.

Case 1. a = 0

Here, $e = a + b = b$.

$$F_1 = \int_0^b f_1(s)q(s)ds \qquad (3.195)$$

$$F_4 = \int_0^b f_4(s)q(s)ds \qquad (3.196)$$

Case 2. c = 0

Here, $a + b = L_3$ and $e = L_3$.

$$F_1 = \int_a^e f_1(s)q(s)ds \qquad (3.197)$$

$$F_4 = \int_a^e f_4(s)q(s)ds \qquad (3.198)$$

Case 3. a = 0 and c = 0

$$F_1 = \int_0^e f_1(s)q(s)ds \qquad (3.199)$$

$$F_4 = \int_0^e f_4(s)q(s)ds \qquad (3.200)$$

Thus far in this section, we have dealt with the general situation for two categories: (1) material and section properties and (2) span lengths. Now we can consider special cases where we have constant material and section properties, or constant span lengths for the whole structure, or the combination of uniformity in both categories by directly incorporating the defining properties of each of these cases in the expressions above.

Case 4. $L_1 = L_2 = L_3 = L$

Under this category for the general case where a and c are both nonzero, F_1 and F_4 are given formally by Formulae (3.191) and (3.192) respectively.

If $a = 0$, then F_1 and F_4 are given formally by Formulae (3.195) and (3.196) respectively. If $c = 0$, then F_1 and F_4 are given formally by Formulae (3.197) and (3.198) respectively. If both $a = 0$ and $c = 0$, then F_1 and F_4 are given formally by Formulae (3.199) and (3.200) respectively.

Case 5. $J_1 = J_2 = J_3 = J$

This case can be treated in the same manner.

Case 6. $L_1 = L_2 = L_3 = L$ and $J_1 = J_2 = J_3 = J$

This case can be treated in the same manner.

3.2.3. Uniform Load on an Exterior Span

The beam with its loading is depicted in Figure 3.8. The formal solutions are still given by (3.189) and (3.190) in Section 3.2.2, with Q_1, Q_4, and D given by (3.187), (3.188), and (3.157) respectively in Section 3.2.1. F_1 and F_4 can be obtained from Section 3.2.2 by setting $q(s) = w$, a constant. However, for the sake of experiencing diversified forms of solutions, we will use the notations introduced in Section 3.2.1. The results are given below.

$$R_1 = \frac{h_1 E_2 + h_2 E_3 + h_3 E_4}{D} \tag{3.201}$$

$$R_4 = \frac{k_1 E_2 + k_2 E_3 + k_3 E_4}{D} \tag{3.202}$$

where

$$h_1 = Z\left(T - \frac{C_2 L_3^2}{3}\right) \tag{3.203}$$

$$h_2 = \frac{-C_3 L_3 Z}{2} \tag{3.204}$$

$$h_3 = \frac{ZL_3}{6J_3} \tag{3.205}$$

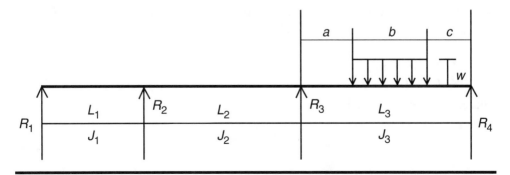

FIGURE 3.8. Uniform load on an exterior span of a three-span continuous beam

$$k_1 = \frac{GC_2L_3}{3} - Z^2L_3 \tag{3.206}$$

$$k_2 = \frac{GC_3}{2} \tag{3.207}$$

$$k_3 = \frac{-G}{6J_3} \tag{3.208}$$

$$E_2 = \frac{e^2 - a^2}{2} \tag{3.209}$$

$$E_3 = \frac{e^3 - a^3}{3} \tag{3.210}$$

$$E_4 = \frac{e^4 - a^4}{4} \tag{3.211}$$

D is as defined in Section 3.2.1. e is as defined in previous sections, namely $e = a + b$.

A list of special cases is given below with simplified results.

Case 1. a = 0

Here, $e = b$.

$$E_2 = \frac{b^2}{2} \tag{3.212}$$

$$E_3 = \frac{b^3}{3} \tag{3.213}$$

$$E_4 = \frac{b^4}{4} \tag{3.214}$$

R_1 and R_4 are given formally by (3.201) and (3.202) respectively, with E_2, E_3, and E_4 given by (3.212–3.214) respectively.

Case 2. c = 0

This means that $e = L_3$. Hence

$$E_j = \frac{L_3^j - a^j}{j}, \quad j = 2, 3, 4 \tag{3.215}$$

The formal solutions R_1 and R_4 are given by (3.201) and (3.202) respectively, with simplified E_j ($j = 2, 3, 4$) given by (3.215).

Case 3. *a* = *c* = 0

This means that $b = L_3$. Hence, $e = L_3$ and

$$E_j = \frac{L_3}{j}, \quad j = 2, 3, 4 \tag{3.216}$$

The formal solutions R_1 and R_4 are given by (3.201) and (3.202) respectively, with simplified E_j ($j = 2, 3, 4$) given by (3.216).

We have treated the most general case of a uniform load on an exterior span. Let us look at the special cases by following the same format of presentation as in Section 3.2.2.

Case 4. $L_1 = L_2 = L_3 = L$

For the general case under this category, the solutions R_1 and R_4 are given formally by (3.201) and (3.202) respectively, with simplified h_1, h_2, h_3, k_1, k_2, and k_3 given above in this section. This simplification is due to geometry and is not related to loads. We can get the information on the location and extent of load from E_2, E_3, and E_4 given in cases 1, 2, or 3 above as appropriate. That is, we still have the three major special cases regarding the location and extent of load under the category of equal span lengths.

Case 5. $J_1 = J_2 = J_3 = J$

The formal solutions R_1 and R_4 are again given by (3.201) and (3.202) respectively, with simplified h_j, k_j ($j = 1, 2, 3$) obtained as in Section 3.2.1 from setting $J_1 = J_2 = J_3 = J$.

Regarding the effects of location and extent of load on the reactions, exactly the same remarks as given in case 4 can be made here.

Case 6. $L_1 = L_2 = L_3 = L$ and $J_1 = J_2 = J_3 = J$

The formal solutions R_1 and R_4 are still given by (3.201) and (3.202) respectively. We use the appropriate expressions from Section 3.2.1 for all entities involved in h_j, k_j ($j = 1, 2, 3$) in this section in computing h_j, k_j ($j = 1, 2, 3$) and use the remarks in case 4 regarding E_j ($j = 2, 3, 4$).

For the special case when $a = 0 = c$, the results are as follows:

$$R_1 = \frac{wL}{60} \tag{3.217}$$

$$R_2 = \frac{-wL}{10} \tag{3.218}$$

$$R_3 = \frac{13wL}{20} \tag{3.219}$$

$$R_4 = \frac{13wL}{30} \tag{3.220}$$

3.2.4. An Alternative Treatment of the Problem of Uniform Load on an Exterior Span

Consider again the general three-span continuous beam with all hinged supports as shown in Figure 3.8 in Section 3.2.3.

The reactions are

$$R_1 = \left(\frac{Qw}{D}\right)[(f_1)(E_2) + (f_2)(E_3) + (f_3)(E_4)] \tag{3.221}$$

$$R_2 = \frac{(R_4)(L_3) - (R_1)A}{L_2} - \left(\frac{E_2}{L_2}\right)w \tag{3.222}$$

$$R_3 = w\left(b + \frac{E_2}{L_2}\right) + \frac{(R_1)(L_1) - (R_4)B}{L_2} \tag{3.223}$$

$$R_4 = -\left(\frac{w}{D}\right)[(g_1)(E_2) + (g_2)(E_3) + (g_3)(E_4)] \tag{3.224}$$

The bending moments are

$$M_1 = (R_1)x \quad \text{for } 0 < x < L_1 \tag{3.225}$$

$$M_2 = (R_1)x + (R_2)[x - (L_1)] \quad \text{for } L_1 < x < A \tag{3.226}$$

$$M_3 = (R_4)z \quad \text{for } 0 < z < c \tag{3.227}$$

$$M_4 = (R_4)z - \left(\frac{w}{2}\right)(z - c)^2 \quad \text{for } c < z < (b + c) \tag{3.228}$$

$$M_5 = (R_4)z - (wb)\left[z - \frac{b + 2c}{2}\right] \quad \text{for } (b + c) < z < L_3 \tag{3.229}$$

$$\max M = (R_4)\left[c + \frac{R_4}{2w}\right] \quad \text{(for } M_4\text{) at } z = c + \frac{R_4}{w} \tag{3.230}$$

where

$$A = L_1 + L_2 \tag{3.231}$$

$$B = L_2 + L_3 \tag{3.232}$$

$$C_i = \frac{L_i}{J_i}, \quad i = 1, 2, 3 \tag{3.233}$$

$$D = [3(C_2)^2 + 4(C_1)(C_2) + 4(C_2)(C_3) + 4(C_3)(C_1)]\left[(L_1)\frac{L_3}{6}\right]^2 \tag{3.234}$$

$$e = a + b \tag{3.235}$$

$$G = [(C_1) + (C_2)]\frac{(L_1)^2}{3} \tag{3.236}$$

$$H = Q(L_3) = (L_1)(L_2)\frac{L_3}{6(J_2)} \tag{3.237}$$

$$Z = [(C_2) + (C_3)]\left[\frac{(L_3)^2}{3}\right] \tag{3.238}$$

$$E_2 = \frac{e^2 - a^2}{2}, \qquad E_3 = \frac{e^3 - a^3}{3}, \qquad E_4 = \frac{e^4 - a^4}{4} \tag{3.239}$$

$$f_1 = (L_3)^2\frac{C_3}{3}, \qquad f_2 = -(L_3)\frac{C_3}{2}, \qquad f_3 = \frac{L_3}{6(J_3)} \tag{3.240}$$

$$g_1 = (L_3)\left[Q^2G\frac{C_2}{3}\right], \qquad g_2 = -G\frac{C_3}{2}, \qquad g_3 = \frac{G}{6(J_3)} \tag{3.241}$$

$$Q = (L_1)\frac{L_2}{6(J_2)} \tag{3.242}$$

Some special cases are displayed below.

Case 1. a = 0

The results can be simplified as follows:

$$R_2 = \frac{(R_4)(L_3) - (R_1)A}{L_2} - [B^2]\frac{w}{2L_2} \tag{3.222A}$$

$$R_3 = w\left[b + \frac{b^2}{2L_2}\right] + \frac{(R_1)(L_1) - (R_4)b}{L_2} \tag{3.223A}$$

Formula (3.228) remains but with $b + c = L_3$. Formula (3.229) is out. Formula (3.235) becomes $e = b$. Formulae in (3.239) become

$$E_2 = \frac{b^2}{2}, \qquad E_3 = \frac{b^3}{3}, \qquad E_4 = \frac{b^4}{4} \tag{3.239A}$$

The rest of the formulae in the general case remain the same.

Case 2. *c = 0*

The results in the general case can be simplified as follows. Formula (3.227) is out. Formula (3.228) becomes

$$M_4 = (R_4)z - \frac{wz^2}{2} \quad \text{for } 0 < z < b \tag{3.228B}$$

Formula (3.229) becomes

$$M_5 = (R_4)z - wb\left(z - \frac{b}{2}\right) \quad \text{for } b < z < L_3 \tag{3.229B}$$

Formula (3.230) becomes

$$\max M = \frac{R_4}{2w} \quad \text{for } M_4 \text{ at } z = \frac{R_4}{w} \tag{3.230B}$$

All other formulae remain unchanged.

Case 3. *$L_1 = L_2 = L_3 = L$*

The results in the general case can be simplified as follows. Formulae (3.225–3.228) remain the same except here $L_i = L$ ($i = 1, 2, 3$), $A = 2L$, and $b + c = L$.
Formulae (3.230–3.233) become

$$A = 2L \tag{3.230C}$$

$$B = 2L \tag{3.231C}$$

$$C_i = \frac{L}{J_i}, \quad i = 1, 2, 3 \tag{3.232C}$$

$$D = \left\{\frac{3}{(J_2)^2} + \frac{4}{(J_1)(J_2)} + \frac{4}{(J_2)(J_3)} + \frac{4}{(J_3)(J_1)}\right\} \frac{(L)^6}{36} \tag{3.233C}$$

Also, here

$$f_1 = \frac{L^3}{3J_3}, \qquad f_2 = \frac{-L^2}{2(J_3)}, \qquad f_3 = \frac{L}{6(J_3)} \qquad (3.239C)$$

$$g_1 = -\left[\frac{4}{J_1} + \frac{3}{J_2}\right]\frac{L^5}{36(J_2)}, \qquad g_2 = -(L^4)\frac{T}{6(J_3)},$$

$$\hspace{8cm}(3.241C)$$

$$g_3 = (L^3)\frac{T}{18(J_3)}$$

where

$$T = \frac{1}{J_1} + \frac{1}{J_2} \qquad (3.241C')$$

The rest of the formulae remain the same as in the general case.

Let us consider the interesting situation when $a = 0$. Expressions for E_2 and E_3 reduce to $b^2/2$ and $b^3/3$ respectively as in case 1, with the resulting simplified formulae contained therein, in addition to the simplified formulae in case 3 above.

Similarly, the important and interesting situation of $c = 0$ can be viewed in light of the results of cases 2 and 3.

Case 4. $J_1 = J_2 = J_3 = J$

The results in the general case can be reduced as follows. Formulae (3.221–3.238) remain the same, while (3.240) and (3.421) become

$$f_1 = \frac{(L_3)^3}{3J}, \qquad f_2 = -\frac{(L_3)^2}{2J}, \qquad f_3 = \frac{L_3}{6J} \qquad (3.240D)$$

$$g_1 = -(L_1)^2(L_2)(L_3)\frac{4(L_1) + 3(L_2)}{36(J)^2}, \qquad g_2 = -A(L_1)^2\frac{(L_3)}{6J^2},$$

$$\hspace{8cm}(3.241D)$$

$$g_3 = A\frac{(L_1)^2}{18J^2}$$

Again, the two situations $a = 0$ only and $c = 0$ only can be dealt with in the same manner as before.

Case 5. $L_1 = L_2 = L_3 = L$ and $J_1 = J_2 = J_3 = J$

The results in the general case can be further simplified to the following. Formulae (3.221–3.228) remain valid, while (3.240) and (3.241) become

$$f_1 = \frac{L^3}{3J}, \qquad f_2 = \frac{-L^2}{2J}, \qquad f_3 = \frac{L}{6J} \tag{3.240E}$$

$$g_1 = \frac{-7L^5}{(6J)^2}, \qquad g_2 = \frac{-L^4}{3J^2}, \qquad g_3 = \frac{L^3}{9J^2} \tag{3.241E}$$

respectively.

Once again, the important problems arising from setting $a = 0$, $c = 0$, or $a = c = 0$ can be dealt with in exactly the same manner as described above in other parts of this section.

Case 6. a = 0 and c = 0

The results can be further simplified as follows:

$$R_2 = \frac{(R_4)(L_3) - (R_1)A}{L_2} - \frac{(B^2)w}{2(L_2)} \tag{3.222F}$$

$$R_3 = w(L_3)\left\{1 + \frac{L_3}{2(L_2)}\right\} + \frac{(R_1)(L_1) - (R_4)B}{L_2} \tag{3.223F}$$

Formulae (3.227) and (3.229) are out, while Formula (3.228) becomes

$$M_4 = (R_4)z - \frac{wz^2}{2} \quad \text{for } 0 < z < L_3 \tag{3.228F}$$

with

$$\max M_4 = \frac{(R_4)^2}{2w} \quad \text{at } z = \frac{R_4}{w} \tag{3.230F}$$

$$e = b = L_3 \tag{3.234F}$$

$$E_2 = \frac{(L_3)^2}{2}, \qquad E_3 = \frac{(L_3)^3}{3} \tag{3.239F}$$

All other formulae in the general case remain valid here.

Case 7. L₁ = L₂ = L₃ = L and a = 0 = c

The above formulae in case 6 can be reduced even further now. The simplified results are

$$R_2 = [(R_4) - 2(R_1)] - 2wL \tag{3.222G}$$

$$R_3 = 1.5wL + (R_1) - 2(R_4) \tag{3.223G}$$

$$e = b = L \tag{3.234G}$$

$$E_2 = \frac{L^2}{2}, \quad E_3 = \frac{L^3}{3} \tag{3.239G}$$

Formulae (3.228F) and (3.230F) remain unchanged and all other formulae in the general case remain valid here. Furthermore, all the formulae in case 3 remain valid here.

Case 8. $J_1 = J_2 = J_3 = J$ and $a = c = 0$
The results in both cases 4 and 6 apply.

Case 9. $L_1 = L_2 = L_3 = L$, $J_1 = J_2 = J_3 = J$, and $a = c = 0$
The results in both cases 5 and 6 apply. Thus, explicitly, the reactions are

$$R_1 = \frac{wL}{60} \tag{3.221I}$$

$$R_2 = \frac{-wL}{10} \tag{3.222I}$$

$$R_3 = \frac{13wL}{20} \tag{3.223I}$$

$$R_4 = \frac{13wL}{30} \tag{3.224I}$$

The bending moments are given in case 6 above.

Illustrative Example for a General Three-Span Continuous Beam with Uniform Load on an Exterior Span: General Characteristics of the Solution

Consider the general case of a three-span continuous beam with uniformly distributed load over part of an exterior span (see Figure 3.8). This is a second-order statically indeterminate structure. We proceed to apply the method of least work to solve the two unknown reactions from the resulting system of equations. The question here is: Out of the four reactions, which two will we pick as the two unknowns in the equations? The choice will affect the degree of complexity and the extent of manipulative operations needed to get the final answer. Recall that the entire set of solutions is the same no matter which reactions are selected as the unknowns in the system of equations, and we do the naming and labeling of mathematical entities systematically before doing anything else to them. With this, upon choosing R_1 and R_4 as the two unknowns for the system of equations, we obtain R_1 and R_4 as shown in Formulae (3.221) and (3.224) respectively.

The other reactions R_2 and R_3 are obtained from equations of equilibrium and are expressions involving the newly obtained R_1 and R_4 among other things. We look at the solutions obtained and note that the key to a comprehensive account of what is happening in the solution set is to examine the expressions for R_1 and R_4 first and see what conclusions we can draw from the investigation.

Let us look at R_1 now. At first glance, we notice that it is basically a product of two factors. Thus,

$$R_1 = (A_1)(B_1) \tag{3.243}$$

where

$$A_1 = \frac{Qw}{D} \tag{3.244}$$

and

$$B_1 = [(f_1)(E_2) + (f_2)(E_3) + (f_3)(E_4)] \tag{3.245}$$

Using (3.239) and (3.240), we can rewrite

$$B_1 = (L_3)\left[(L_3)(C_3)\frac{E_2}{3} - (C_3)\frac{E_3}{2} + \frac{E_4}{6J_3} \right] \tag{3.246}$$

Of course, we can also say that R_1 is a product of two factors but with the definitions of A_1 and B_1 different from the above, or we can even say that R_1 is a product of three factors, with appropriate definitions of the three factors. It can be seen clearly that there is ample freedom of choice regarding the direction and approach of exploration. It is interesting to pursue any one of these options and reach a set of results which possess certain characteristic features of significant physical applications. What is displayed below demonstrates one particular choice of approach for exploration. Thus, the point is the following simple statement: We can make our own choice in any way we want as long as we are consistent and stick to our choice throughout the course of problem solving or research.

How does R_1 vary with A_1 or B_1? We focus on the definitions of the components of A_1 and B_1 and do our analysis as follows. We now deal with A_1, which is itself a product of two factors: Q and w/D. We note that both Q and D have the same dimension, namely length to the (-2) power. Therefore, A_1 has the same dimension as w. Also, w is harmless and appears in many places. The entity Q, from Formula (3.242), can be rewritten as

$$Q = (L_1)\frac{C_2}{6} \tag{3.247}$$

We observe here that Q has nothing to do with quantities associated with the loaded span. As we can see from (3.246), the entity B_1 has nothing to do with any quantities associated with the unloaded spans.

What about D? Well, it has to do with all spans. For example, one of its factors is a quadratic expression that contains material and section properties of all spans, and let us call it N. Moreover, we spot that D appears in R_4 also. As a matter of fact, D appears in R_2 and R_3 as well. Thus, it is critical to get a handle on the behavior of D before we call upon it to act together with other players and assess the effects of all these players on R_1.

We will make it our project to treat D by identifying its components with the defining characteristics and to study the effects on D of varying one parameter while holding the rest constant. For convenience, let's rewrite the defining formula for D as shown in (3.234) in the following manner

$$D = N[(L_1)(L_3)]^2 \tag{3.248}$$

where

$$N = \frac{3(C_2)^2 + 4(C_1)(C_2) + 4(C_2)(C_3) + 4(C_3)(C_1)}{36} \tag{3.249}$$

Note that D involves L_i and J_i $(i = 1, 2, 3)$ as fundamental independent variables after replacing C_i, which appears in N, by the definition $C_i = L_i/J_i$ $(i = 1, 2, 3)$, as given in (3.233).

It is clear that, among many possible ways to do things, a spreadsheet approach will do the job.

Having completed a brief description of A_1, let's do B_1. First, we rewrite R_1 as

$$R_1 = w\left(\frac{Q}{D}\right)(L_3)X \tag{3.250}$$

where

$$X = (L_3)(C_3)\frac{E_2}{3} - (C_3)\frac{E_3}{2} + \frac{E_4}{6J_3} \tag{3.251}$$

Note that

$$B_1 = (L_3)X \tag{3.252}$$

The entity X involves the location and extent of the distributed load expressed in E_j $(j = 2, 3, 4)$ as well as the material and section properties and length of the span in which the load is applied and nothing else. In other words, X has nothing to do with the other two spans. X appears to be a function of L_3, J_3, and C_3, when E_j $(j = 2, 3, 4)$ are held as constants. However, from Formula (3.233), under this condition, X is a function of L_3 and J_3, the fundamental independent variables.

We can take care of the theme on R_1 in seven steps, as follows.

Step 1. First, we notice that L_3 appears in many places in (3.251) and in fact xX is our old friend B_1 (when we call L_3 as x), which we met in (3.252). As before, if we want to see the effects of varying x on R_1, we can hold, in R_1, all the entities other than x as constants and consider R_1 as a function of x only. The strategy now is clear from (3.251) and (3.252) above: Holding E_j ($j = 2, 3, 4$) and J_3 constant, then X is a function of x only. Thus, we have

$$R_1 = wx\left(\frac{Q}{D}\right)X \tag{3.253}$$

Here X is a second-degree polynomial in x. However, X is not the sole expression containing x, not even xX is, as can be seen in (3.255) below. The idea is once again that we have to look at the whole picture. In other words, we must see both the trees and the forest.

We rewrite

$$R_1 = YxwX \tag{3.254}$$

where

$$Y = \frac{Q}{D} = 6\frac{C_2}{(L_1)nx^2} \tag{3.255}$$

and n is a first-degree polynomial in x, resulting from holding all other entities in N as constants.

Therefore,

$$R_1 = (S_1)f \tag{3.256}$$

where

$$f = \frac{X}{nx} \tag{3.257}$$

We observe that f is a ratio of two quadratic polynomials in x and collects everything that involves x, while S_1 is a coefficient independent of x. Our task now is to investigate f. Again, there are many ways to do the job, by computer alone. Of course, we can also tackle the problem by analytic means. In fact, for the sake of actively experiencing a much deeper insight into the subject matter, it is worthwhile to conduct an analytic investigation as outlined below. Note that this approach has the versatility to accommodate simple computer work at intermediate steps of the analytical investigation to suit our particular needs and interest. Thus, we will proceed in this manner.

Now, for convenience and efficiency of operations, we will write

$$X = x^2(K_2) + x(K_1) + (K_0) \qquad (3.258)$$

where K_2, K_1, and K_0 are independent of x.

We write

$$xn = x^2(d_2) + x(d_1) \qquad (3.259)$$

where d_2 and d_1 are independent of x.

Once again, we can proceed in many ways. One approach is to go directly to our spreadsheets and get the answers. Another way is to use the analytical approach at first and see what results we can get out of this investigation. We get, after keeping our packages intact during the process,

$$f = (h_0) + \frac{h_1}{x} \qquad (3.260)$$

where h_0 and h_1 are independent of x.

From (3.260), it is convenient to start to use spreadsheets if we wish. Let's go one step further along the analytic road by noting that f is the sum of two terms. One of them is a constant and the other is a hyperbola in x.

Looking at this in light of (3.256), we see immediately from another point of view that R_1 is made up of the sum of two ingredients. One is inversely proportional to x, and the other is a constant when the length of the loaded span is the one and only independent variable.

Step 2. Next, to see the effects of C_2 on the reaction R_1, let us call C_2 as y for convenience and look at Formula (3.254). Realizing that x, w, and X are all independent of y, we see that y is the parameter that we want to study. Further examination reveals that we need to consider specifically y/N. Under the conditions just stated, N is a quadratic polynomial function of y. Thus, our task boils down to investigating a function F of the form

$$f = \frac{y}{(N_2)y^2 + (N_1)y + (N_0)} \qquad (3.261)$$

There are many ways to tackle this problem. Let us look at one method outlined below.

Using the standard method in calculus, let's find the first derivative of this function. First, we observe that all the coefficients are positive. Then we rewrite Expression (3.261) as

$$f = \left[\frac{1}{N_2} \right] g \qquad (3.262)$$

where

$$g = \frac{y}{h}, \qquad h = [y^2 + y(b_1) + (b_0)] \tag{3.263}$$

Now we concentrate on g as the function to be studied. We obtain

$$g' = \frac{(b_0) - y^2}{(h)^2} \tag{3.264}$$

We see that depending on whether $[(b_0) - y^2]$ is negative or positive, g is a decreasing or increasing function of y. When $g' = 0$, the test for extreme values of g yields the result that

$$y = (b_0)^{\frac{1}{2}} \tag{3.265}$$

is the only solution by noting that a negative value of y is physically not admissible. Further test indicates that we have a maximum value of g at the critical point defined by (3.265). Thus, we learn a great deal about the function g.

Let's go back to f defined by (3.262) and find out how f varies as y varies through the auxiliary variable g. In other words, we have a firm grasp of the behavior of C_2 regarding how it can affect R_1 analytically.

Step 3. If we want to see the effects of varying C_1 on R_1, we need to study the function $1/N$ mainly. Furthermore, here N is a much simplified expression when we treat it as a function of C_1. Let's call C_1 as z for convenience. Thus,

$$N = (q_1)z + (q_0) \tag{3.266}$$

where q_1 and q_0 are positive quantities which are independent of z.

Let

$$f = \frac{1}{N} \tag{3.267}$$

and f is the function to be examined. We observe that f is a monotone decreasing function of z, and we can easily plot a graph or do other displays to depict the behavior of f and consequently that of R_1.

Step 4. If we want to see what C_3 can do to influence R_1, we can look at X/N. Here, again it is understood that both X and N are treated as functions of C_3 only. We may call C_3 by any name under this discussion, so let us call it z. The function to be studied is

$$f = \frac{X}{N} \tag{3.268}$$

where

$$X = (P_1)z + (P_0) \tag{3.269}$$

and

$$N = (Q_1)z + (Q_0) \qquad (3.270)$$

This time, we have f as a quotient of two first-degree polynomials in the independent variable z. After a straightforward manipulation, we can reach

$$f = \frac{F_1}{z + (D_0)} + (F_0) \qquad (3.271)$$

where D_0, F_0, and F_1 are independent of z.

We have seen this kind of function earlier. It is one of our old friends, and we know how to take care of the packages and what the symbols F_1 and F_0 stand for in terms of the given information on the structure. Thus far, we have treated C_i as the independent variable for a specific i under consideration if we choose not to open up the package C_i for that i. Note that $C_i = L_i/J_i$, where $i = 1, 2, 3$. Thus, we can proceed in one of two ways: either keep one of the two entities here constant while varying the other or vary both L_i and J_i. We see that C_i is actually an intermediate variable. In practical applications, we can get results in either way, depending on the need and objectives of the investigation.

Let's count the fundamental independent variables in R_1. We have the list: L_i, J_i ($i = 1, 2, 3$), w, e, and a. Note that E_j ($j = 2, 3, 4$) and C_i ($i = 1, 2, 3$) are intermediate variables. Let us continue the exploration.

Step 5. We look at the effects of L_1, and call L_1 as u for convenience. Consider the function $1/U$ which will be appearing in R_1.

$$R_1 = \frac{6(C_2) \dfrac{wX}{L_3}}{U} \qquad (3.272)$$

where

$$U = [(G_1)u + (G_0)]u \qquad (3.273)$$

and G_0 and G_1 are entities which are independent of u.

In the expression for R_1, the independent variable is u, which is contained entirely in U. Therefore, the effects of changing u are reflected in U. We can see that R_1 is inversely proportional to U. Now, by investigating U as a function of u by the standard method in calculus, we see that U is an increasing function of u. Also, the coefficients G_0 and G_1 are positive. Of course, we may consider the function $1/U$ instead of U in the above and reach the same conclusion. This shows once again the versatility of approaches available to us for exploration.

Step 6. What about L_2? When we are doing C_2, we know how to handle a function of y in the form

$$\frac{y}{(N_2)y^2 + (N_1)y + N_0}$$ (3.274)

When dealing with L_2 as the independent variable, we have the same kind of function. Therefore, mathematically, the problem-solving method can be the same for both cases.

Step 7. Regarding E_j ($j = 2, 3, 4$), we can view each one as a function of two variables, e and a, and treat it analytically or numerically from that point on. Thus, information about how these entities E_j ($j = 2, 3, 4$) change as a result of varying one of the two independent variables (e, a) can be gathered systematically. We can also treat the problem in the following way. Consider the effects of one parameter at a time in order to get a feel for the whole situation represented by our old friend X. For example, to see the effects of the independent variable e, we will hold all the other entities constant and examine the resulting function X of the distance e. Once again, a spreadsheet approach will be fine.

Commentary

We treated R_1 in detail above. We can treat R_4 in a similar manner. The lesson from looking at the form of R_1 and R_4 is that they have something in common, namely w/D. Looking at $R_4 = (A_4)(B_4)$, where $A_4 = -(w/D)$, $B_4 = [(g_1)(E_2) + (g_2)(E_3) + (g_3)(E_4)]$, we see that A_4 is simpler than A_1, while B_4 appears to be about the same as B_1 in complexity of form, as seen from (3.221), (3.224), (3.240), and (3.241). However, the defining formulae for g_i ($i = 1, 2, 3$) are more complicated than the corresponding f_i ($i = 1, 2, 3$) after viewing Formulae (3.240) and (3.241). The principle is again, depending on our choice of definitions of the ingredients, that the entities with which we are working may have different appearances or "looks." Thus, we can, if we want, rewrite (3.224) in a different form, but we do not have to.

3.2.5. Triangular Load on an Exterior Span

For the general problem with unequal span lengths L_i as well as unequal material and section properties J_i ($i = 1, 2, 3$), the three-span continuous beam under a triangular load on an exterior span is depicted in Figure 3.9.

The triangular load is represented as

$$q(s) = q_0 + q_1 s$$ (3.275)

with

$$q_0 = \frac{-aw}{b}$$ (3.276)

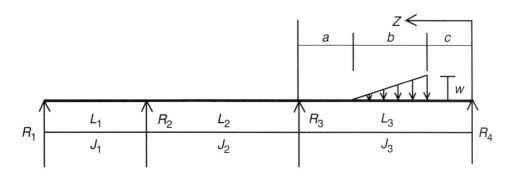

FIGURE 3.9. Triangular load on an exterior span of a three-span continuous beam

$$q_1 = \frac{w}{b} \tag{3.277}$$

The formal solutions are

$$R_1 = Q\,\frac{f_1E_2 + f_2E_3 + f_3E_4 + f_4E_5}{D} \tag{3.278}$$

$$R_4 = \frac{g_1E_2 + g_2E_3 + g_3E_4 + g_4E_5}{D} \tag{3.279}$$

where

$$E_j = \frac{e^j - a^j}{j}, \quad j = 2, 3, 4, 5 \tag{3.280}$$

$$f_1 = q_0A_1 \tag{3.281}$$

$$f_2 = q_0A_2 + q_1A_1 \tag{3.282}$$

$$f_3 = q_0A_3 + q_1A_2 \tag{3.283}$$

$$f_4 = q_1A_3 \tag{3.284}$$

$$g_1 = q_0D_1 \tag{3.285}$$

$$g_2 = q_1D_1 + q_0D_2 \tag{3.286}$$

$$g_3 = q_1D_2 + q_0D_3 \tag{3.287}$$

$$g_4 = q_1D_3 \tag{3.288}$$

with

$$A_1 = \frac{C_3L_3^2}{3} \tag{3.289}$$

$$A_2 = \frac{-C_3 L_3}{2} \tag{3.290}$$

$$A_3 = \frac{C_3}{6} \tag{3.291}$$

$$D_1 = Q^2 L_3 - \frac{C_2 L_3 G}{3} \tag{3.292}$$

$$D_2 = \frac{-C_3 G}{2} \tag{3.293}$$

$$D_3 = \frac{G}{6 J_3} \tag{3.294}$$

$$Q = \frac{L_1 C_2}{6} \tag{3.295}$$

The typical special cases that we examined for other topics can be treated in exactly the same manner as before and are therefore left to the reader as exercises for practice.

3.2.6. A Concentrated Couple at an Arbitrary Point on an Exterior Span

For the general problem with unequal span lengths L_i as well as unequal material and section properties J_i ($i = 1, 2, 3$), the three-span continuous beam under a concentrated couple acting at an arbitrary point on an exterior span is shown in Figure 3.10.

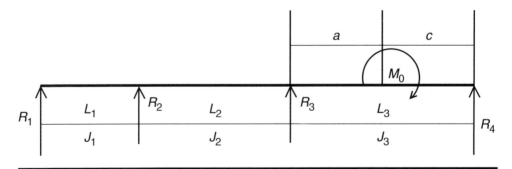

FIGURE 3.10. A concentrated couple at an arbitrary point on an exterior span of a three-span continuous beam

The unknowns of the system of equations resulting from the method of least work are R_1 and R_4 as usual. The solutions to the system of equations are

$$R_1 = \frac{HK - FZ}{D} \tag{3.296}$$

$$R_4 = \frac{FN - GK}{D} \tag{3.297}$$

where

$$F = \frac{-L_1 C_2 M_0}{6} \tag{3.298}$$

$$G = L_1^2 \frac{C_1 + C_2}{3} \tag{3.299}$$

$$H = L_1 L_2 \frac{L_3}{6J_2} = L_1 L_2 \frac{C_2}{6} \tag{3.300}$$

$$N = H \tag{3.301}$$

$$K = -M_0 \frac{2C_3 L_2 + 3\frac{L_3^2 - c^2}{J_3}}{6} \tag{3.302}$$

$$Z = L_3^2 \frac{C_2 + C_3}{3} \tag{3.303}$$

$$D = L_1^2 L_3^2 \frac{3C_2^2 + 4(C_1 + C_3)C_2 + 4C_1 C_3}{36} \tag{3.304}$$

$$C_i = \frac{L_i}{J_i}, \quad i = 1, 2, 3 \tag{3.305}$$

Note that, on the one hand, the definitions of C_i, G, H, and D, which are related to material and geometry only, are as before for the other load cases of the three-span continuous beams, as they should be. On the other hand, the entities F and K are directly related to the load. It is interesting to note also that the effect of the location of the load, namely c, has similar characteristics as in the case of a two-span continuous beam.

For the special cases of importance and interest under the same categories as before, we can have exactly the same simplifications in the expressions G, H, Z, D, and C_i ($i = 1, 2, 3$).

The expressions F and K will have their own simplified versions as a result of applying the attributes of the specific case under consideration as follows. For the special case of equal span lengths, we have

$$F = -\frac{M_0 L^2}{6 J_2} \tag{3.306}$$

$$K = \frac{-M_0(5L^2 - 3c^2)}{6 J_3} \tag{3.307}$$

For the special case of constant material and section properties for all spans, we have

$$F = \frac{-M_0 L_1 L_2}{6 J} \tag{3.308}$$

$$K = \frac{-M_0[2 L_2 L_3 + 3(L_3^2 - c^2)]}{6 J} \tag{3.309}$$

For the special case of equal span lengths as well as constant material and section properties throughout, we have

$$F = -\frac{M_0 L^2}{6 J} \tag{3.310}$$

$$K = \frac{-M_0(5L^2 - 3c^2)}{6 J} \tag{3.311}$$

Moreover, for this last case, we have

$$R_1 = -M_0 L^3 \frac{L^2 - 3c^2}{36 J^2 D} \tag{3.312}$$

$$R_4 = \frac{M_0 L^3 (19 L^2 - 12 c^2)}{36 J^2 D} \tag{3.313}$$

Looking at (3.312) and (3.313), we note that both R_1 and R_4 are proportional to

$$V = \frac{M_0 L^3}{36 J^2 D} \tag{3.314}$$

R_4 and M_0 always have the same sign. However, R_1 and M_0 have the same sign only if $L^2 - 3c^2 < 0$, which means $L^2 < 3c^2$. In fact, for $M_0 > 0$, R_1 may be negative, positive, or even zero, depending on whether $L^2 - 3c^2 > 0$, $L^2 - 3c^2 < 0$, or $L^2 - 3c^2 = 0$ respectively. Some interesting special values of R_1 and R_4 are as follows:

$$c = 0, \qquad R_1 = -VL^2, \qquad R_4 = 19VL^2 \qquad (3.315)$$

$$c = L, \qquad R_1 = 2VL^2, \qquad R_4 = 7VL^2 \qquad (3.316)$$

3.2.7. A Concentrated Force at an Arbitrary Point on the Interior Span

For the general problem with unequal span lengths L_i as well as unequal material and section properties J_i (i = 1, 2, 3), the three-span continuous beam under a concentrated force on the interior span is shown in Figure 3.11. In order to illustrate different treatments of the solution results and raise interest on the part of the reader, the solution to this problem will be presented in the following manner. As usual, we take R_1 and R_4 as the two unknowns in the two simultaneous equations resulting from the method of least work. We have

$$R_1 = VC(A_0 + sA_1)P \qquad (3.317)$$

where

$$V = s \frac{L_2 - s}{6L_2 J_2} \qquad (3.318)$$

$$C = \frac{L_1 L_3^2}{6D} \qquad (3.319)$$

with D defined as before.

$$A_0 = -2L_2 C_3 \qquad (3.320)$$

$$-A_1 = (3C_2 + 2C_3) \qquad (3.321)$$

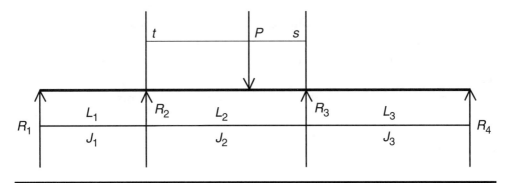

FIGURE 3.11. A concentrated force at an arbitrary point on the interior span of a three-span continuous beam

and

$$R_4 = VC'(D_0 + sD_1)P \tag{3.322}$$

where

$$C' = C\left(\frac{L_1}{L_3}\right) \tag{3.323}$$

$$D_0 = -L_2(4C_1 + 3C_2) \tag{3.324}$$

$$D_1 = 2C_1 + 3C_2 \tag{3.325}$$

with C_i defined as before:

$$C_i = \frac{L_i}{J_i}, \quad i = 1, 2, 3 \tag{3.326}$$

The other reactions are obtained by equations of equilibrium as follows:

$$R_2 = \frac{R_4 L_3 - R_1 A + sP}{L_2} \tag{3.327}$$

$$R_3 = \frac{Pt}{L_2} - \frac{BR_4 - R_1 L_1}{L_2} \tag{3.328}$$

where

$$A = L_1 + L_2, \quad B = L_2 + L_3 \tag{3.329}$$

The bending moments and shear forces can be written down easily using the reactions obtained above and are left to the reader as exercises.

Formulae (3.317–3.321) indicate that once again R_1 is of the same general form as the corresponding solution for the case where the concentrated force is on an exterior span. Moreover, R_1 is a third-degree polynomial in s, and it is basically the product of s and a quadratic expression in s. We have seen this same general characteristic before in the case of a concentrated force on an exterior span. The same remark can be made about R_4.

Special cases of interest with simplified results are as follows.

Case 1. $L_1 = L_2 = L_3 = L$

$$V = s\frac{L - s}{6LJ_2} \tag{3.330}$$

$$C = \frac{L^3}{6D} \tag{3.331}$$

$$A_0 = -LC_3 \tag{3.332}$$

$$A_1 = -(3C_2 + 2C_3) \tag{3.333}$$

$$C_i = \frac{L}{J_i}, \quad i - 1, 2, 3 \tag{3.334}$$

$$C' = C \tag{3.335}$$

$$D_0 = -L(4C_1 + 3C_2) \tag{3.336}$$

$$D_1 = 2C_1 + 3C_2 \tag{3.337}$$

Case 2. $J_1 = J_2 = J_3 = J$

$$V = \frac{s(L_2 - s)}{6L_2 J} \tag{3.338}$$

Formulae (3.317–3.326) are simplified mainly due to the reduced

$$C_i = \frac{L_i}{J} \tag{3.339}$$

Case 3. $L_1 = L_2 = L_3 = L$ and $J_1 = J_2 = J_3 = J$

Formulae (3.330) in case 1 and (3.338) in case 2 are further simplified to

$$V = s\frac{L - s}{6LJ} \tag{3.340}$$

Formulae (3.332), (3.333), (3.336), and (3.337) in case 1 are further simplified because of

$$C_i = \frac{L}{J}, \quad i = 1, 2, 3 \tag{3.341}$$

Thus,

$$A_0 = \frac{-2L^2}{J} \tag{3.342}$$

$$A_1 = \frac{-5L}{J} \tag{3.343}$$

$$D_0 = \frac{-7L^2}{J} \tag{3.344}$$

$$D_1 = \frac{5L}{J} \tag{3.345}$$

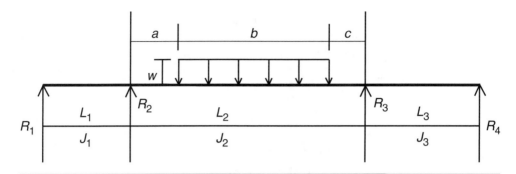

FIGURE 3.12. Uniform load on the interior span of a three-span continuous beam

3.2.8 Uniform Load on the Interior Span

The general three-span continuous beam with unequal span lengths and unequal material and section properties under a uniform load on the interior span is shown in Figure 3.12. The solutions can be obtained by integration of the results in the previous section. We have

$$R_1 = wSL_3Z_1 \tag{3.346}$$

$$R_4 = wSL_1Z_4 \tag{3.347}$$

where

$$S = \frac{L_1L_3}{36J_2L_2D} \tag{3.348}$$

$$Z_1 = (L_2A_0 + A_1L_2^2)E_2 - (2A_1L_2 + A_0)E_3 + A_1E_4 \tag{3.349}$$

$$Z_4 = (L_2D_0 + D_1L_2^2)E_2 - (2D_1L_2 + D_0)E_3 + D_1E_4 \tag{3.350}$$

with A_0, A_1, D_0, and D_1 defined exactly the same as in the case of concentrated force earlier and E_j defined in the usual standard way.

The other reactions are again obtained from the equations of equilibrium as

$$R_2 = \frac{R_4L_3 - R_1A + wb\left(\dfrac{b}{2} + c\right)}{L_2} \tag{3.351}$$

$$R_3 = \frac{wE_2}{L_2} - \frac{BR_4 - L_1R_1}{L_2} \tag{3.352}$$

Some special cases of interest are as follows.

Case 1. a = 0

Then $b + c = L_2$ and $e = b$. Thus, we have simplified

$$E_j = \frac{b_j}{j}, \quad j = 2, 3, 4 \tag{3.353}$$

and simplified Z_1 and Z_4 because of this.

Case 2. c = 0

This means that $a + b = L_2$; thus $e = L_2$. Again, only E_j are affected:

$$E_j = \frac{L_2^j - a^i}{j}, \quad j = 2, 3, 4 \tag{3.354}$$

Case 3. a = 0 and c = 0

This means that the entire interior span is under uniform load. Thus, $e = b = L_2$ and

$$E_j = \frac{L_2^j}{j}, \quad j = 2, 3, 4 \tag{3.355}$$

In the above special cases, Formulae (3.346–3.352) hold with simplified E_j for the individual case at hand to be used in expressions Z_1 and Z_4.

Case 4. Equal Span Lengths

From the above discussion for the general case, we can obtain the solution for the special case of equal span lengths by letting $L_1 = L_2 = L_3 = L$ in the formulae. The formal solution remains the same as in the general case, but with simplified expressions for the components. Thus, we have

$$R_1 = wSLZ_1 \tag{3.356}$$

$$R_4 = wSLZ_4 \tag{3.357}$$

$$R_2 = R_4 - 2R_1 + wb\,\frac{\dfrac{b}{2} + c}{L} \tag{3.358}$$

$$R_3 = \frac{wE_2}{L} - 2R_4 + R_1 \tag{3.359}$$

$$S = \frac{L}{36 J_2 D} \tag{3.360}$$

$$Z_1 = (LA_0 + A_1 L^2)E_2 - (2A_1 L + A_0)E_3 + A_1 E_4 \tag{3.361}$$

$$Z_4 = (LD_0 + D_1 L^2)E_2 - (2D_1 L + D_0)E_3 + D_1 E_4 \tag{3.362}$$

The three important special cases where $a = 0$ only, $c = 0$ only, and both a and c are zero are solved by incorporating the corresponding simplified E_j ($j = 2, 3, 4$) into all the formulae affected. Let us take, for example, the special case where the entire interior span is under uniform load. Here

$$E_j = \frac{L^j}{j}, \quad j = 2, 3, 4 \tag{3.363}$$

$$Z_1 = -(C_2 + 2C_3)\frac{L^4}{4} \tag{3.364}$$

$$Z_4 = -(2C_1 + C_2)\frac{L^4}{4} \tag{3.365}$$

Case 5. $J_1 = J_2 = J_3 = J$

Simplified expressions S, Z_1, and Z_4 are as follows:

$$S = \frac{L_1 L_3}{36 J L_2 D} \tag{3.366}$$

Z_1 and Z_4 are simplified because they contain the reduced expressions for A_0, A_1, A_2, D_0, D_1, and D_2 as a result of the constant product of Young's modulus and section modulus for the entire structure.

Case 6. $L_1 = L_2 = L_3 = L$ and $J_1 = J_2 = J_3 = J$

Further simplification of the expressions for S, Z_1, and Z_4 can be obtained as follows:

$$S = \frac{L}{36 J D} \tag{3.367}$$

$$Z_1 = L(A_0 + A_1 L)E_2 - (A_0 + 2A_1 L)E_3 + A_1 E_4 \tag{3.368}$$

$$Z_4 = L(D_0 + D_1 L)E_2 - (D_0 + 2D_1 L)E_3 + D_1 E_4 \tag{3.369}$$

with the simplified

$$A_0 = \frac{-2L^2}{J} \tag{3.370}$$

$$A_1 = \frac{-5L}{J} \tag{3.371}$$

$$D_0 = \frac{-7L^2}{J} \tag{3.372}$$

$$D_1 = \frac{5L}{J} \tag{3.373}$$

As usual, the special cases where $a = 0$ only, $c = 0$ only, and $a = c = 0$ in cases 5 and 6 can be dealt with in the same manner as before and are left to the interested reader to pursue.

3.2.9. Triangular Load on the Interior Span

For the general problem with unequal span lengths L_i as well as unequal material and section properties J_i ($i = 1, 2, 3$), the three-span continuous beam under a triangular load on the interior span is shown in Figure 3.13.

In order to illustrate the diversity of approaches to solve similar problems, let us use the following notations. For the same entity V encountered in the case with a concentrated force, we have

$$V = sV_1 + s^2V_2 \tag{3.374}$$

where

$$V_1 = \frac{1}{6J_2}, \qquad V_2 = \frac{-1}{6L_2J_2} \tag{3.375}$$

The load intensity function is

$$q = q_0 + sq_1 \tag{3.376}$$

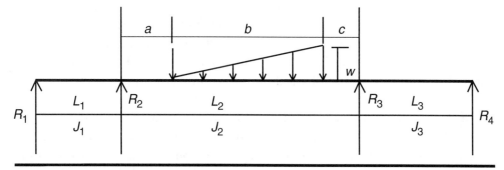

FIGURE 3.13. Triangular load on the interior span of a three-span continuous beam

with

$$q_0 = -\frac{aw}{b}, \qquad q_1 = \frac{w}{b} \tag{3.377}$$

Then

$$qV = sZ_1 + s^2 Z_2 + s^3 Z_3 \tag{3.378}$$

where

$$Z_1 = q_0 V_1 = -aU \tag{3.379}$$

$$Z_2 = q_1 V_1 + q_0 V_2 = \left(1 + \frac{a}{L_2}\right)U \tag{3.380}$$

$$Z_3 = q_1 V_2 = \frac{-U}{L_2} \tag{3.381}$$

with

$$U = \frac{w}{6J_2 b} \tag{3.382}$$

Thus,

$$R_1 = C[A_0 Z_1 E_2 + (A_0 Z_2 + A_1 Z_1)E_3 + (A_0 Z_3 + A_1 Z_2)E_4 + (A_1 Z_3)E_5] \tag{3.383}$$
$$R_4 = C'[D_0 Z_1 E_2 + (D_0 Z_2 + D_1 Z_1)E_3 + (D_0 Z_3 + D_1 Z_2)E_4 + (D_1 Z_3)E_5] \tag{3.384}$$

where

$$C' = C\left(\frac{L_1}{L_3}\right) \tag{3.385}$$

and A_0, A_1, D_0, D_1, and D are as defined in Section 3.2.7; E_j ($j = 2, 3, 4, 5$) is defined in the usual way as before.

We see immediately that the expressions for R_1 and R_4 are very similar, because of the approach using the generic problem, among other things.

It is interesting to note that Expression (3.378) lumps all the entities related to load and that the solution for R_1 just comes out in the wash after we perform integration over s, the result of which is shown within the brackets in (3.383). A similar remark can be made for R_4.

The notations Z_1, Z_2, and Z_3 in this section have nothing to do with the Z's in Section 3.2.8. Of course, we could use different symbols even in different sections.

Regarding the special cases, we can handle them in the same manner as in the previous sections. Even there, they vary somewhat from problem to problem. For the sake of demonstrating the diversity of approaches, we can define a four-by-four matrix with the following four categories: (1) general span lengths and general material and section properties (i.e., the lengths as well as the material and section properties for the three spans are different), (2) only the span lengths for the three spans are the same, (3) only the products or material and section properties for the three spans are the same, and (4) the three spans all have the same length as well as the product of material and section properties. Under each of these categories there are four cases: (a) distances a and c are both nonzero, (b) $a = 0$ only, (c) $c = 0$ only, and (d) $a = 0$ and $c = 0$. It is left to the interested reader to fill in the details of the layout of his or her own choice.

4

BEAMS ON ELASTIC FOUNDATIONS

4.1. BEAMS OF INFINITE LENGTH

4.1.1. A Concentrated Force on the Beam

A concentrated force of magnitude P acts on an elastic beam on elastic foundations of infinite length, with elastic modulus k.

The deflection is given by

$$Y = CG[\cos by + \sin bx] \qquad (4.1)$$

where $C = P/(8b^3J)$ is a constant and

$$G = \exp(-bx) \qquad (4.2)$$

$$b = \left(\frac{k}{4J}\right)^{1/4} \qquad (4.3)$$

$$J = EI_z \qquad (4.4)$$

with E = Young's modulus and I_z = moment of inertia of the beam of constant cross section with respect to the horizontal principal axis z.

We want to see the effects of varying b on the deflection. Therefore, we treat Y as a function of b only while holding all other entities as constants. Thus, we take the first partial derivative of Y, denoted by Y' purely for convenience, with respect to b and obtain

$$Y' = -2CGx \sin bx + A \qquad (4.5)$$

where

$$A = -3Pb^{-4}G\,\frac{\cos bx\,+\,\sin bx}{8J} \tag{4.6}$$

Note that from Expressions (4.5) and (4.6) above

$$\sin bx > 0 \quad \text{and} \quad \cos bx > 0 \tag{4.7}$$

together constitute a set of sufficient conditions for $Y' < 0$. This means that for a fixed x and for the values of b taken from the domain of definition satisfying (4.7), we have Y as a decreasing function of b. By the same token, the simultaneous satisfaction of

$$\sin bx < 0 \quad \text{and} \quad \cos bx < 0 \tag{4.7A}$$

constitutes a set of sufficient conditions for $Y' > 0$. Remarks similar to those made regarding Y as a decreasing function of b above can be made concerning Y as an increasing function of b.

We observe from the expressions for Y and Y' that both of them contain G and are proportional to G. Regarding Y, it is proportional to two additional factors, namely C and f, where f stands for $(\sin bx + \cos bx)$, to which we will refer frequently below.

Now, can we say something about the sign of Y and the possibility of Y being zero?

◆ $Y = 0$ when $f = 0$, which means $\tan bx = 1$, or equivalently $bx = 3\pi/4 + n\pi$.
◆ $Y < 0$ or $Y > 0$ when $f < 0$ or $f > 0$ respectively, which means $\tan bx < 0$ or $bx > 0$ respectively. Thus, f is an indicator regarding Y being positive or negative.

It is interesting to view Y from a rather unusual perspective, namely from looking for an upper bound for Y regardless of the actual b, x values. Hence, consider f as a function of b. The function f has a maximum value of $2^{\frac{1}{2}}$ when $\tan bx = 1$. Thus, Y has an upper bound $2^{\frac{1}{2}}CG$. Therefore, f serves not only as an indicator mentioned above, but also as a scale for an upper bound for Y.

Let us examine the behavior of CG as a function of b only. We see that CG, as a decreasing function of b, has an upper bound C because $G < 1$ except when $bx = 0$, in which case $G = 1$. For C to be meaningful practically, b should be nonzero. We can see that both $2^{\frac{1}{2}}C$ and $2^{\frac{1}{2}}CG$ can serve as upper bounds for Y, but CG is a more precise upper bound for all nonzero x values. Of course, for $x = 0$, we have $CG = C$.

Expression (4.5) can be rewritten as

$$Y' = \frac{-PGN}{2k} \tag{4.8}$$

where

$$N = (2bx + 3) \sin bx + 3\cos bx \qquad (4.9)$$

Note that from (4.8), $Y' = 0$ when $G = 0$ or $N = 0$. The former means that bx is approaching infinity, and for finite b in usual cases of practical applications this implies that the point of interest is at infinity where $Y = 0$ regardless of b, as can be seen from the expression for Y. This leads to

$$\tan bx = \frac{-3}{3 + 2bx} \qquad (4.10)$$

which corresponds to $N = 0$ mentioned above and which covers two cases:

◆ Y has a minimum value when $\cos bx > 0$, with (4.10) satisfied.
◆ Y has a maximum value when $\cos bx < 0$, with (4.10) holding true.

From (4.8), we see that N determines the sign of $Y': Y' > 0$ when $N < 0$ and vice versa. Thus, from (4.9), we have $Y' < 0$ when

$$\tan bx > \frac{-3}{3 + 2bx} \qquad (4.11)$$

which specifies the condition for a decreasing Y.
$Y' > 0$ when

$$\tan bx < \frac{-3}{3 + 2bx} \qquad (4.12)$$

which describes the condition for an increasing Y.

Next, let us consider Expression (4.13) for the bending moment as a function of b only:

$$M = -PG \frac{\sin bx - \cos bx}{4b} \qquad (4.13)$$

Looking at (4.13), we see that $M < 0, = 0,$ or > 0 when $g < 0, = 0,$ or > 0 respectively, meaning when $\tan bx < 0, = 0,$ or > 0 respectively.
We have

$$M' = \frac{PGD}{4} \qquad (4.14)$$

where

$$D = (ag - f) \qquad (4.15)$$

with

$$a = \frac{bx + 1}{b^2} \tag{4.16}$$

$$f = \cos bx + \sin bx \tag{4.17}$$

$$g = \sin bx - \cos bx \tag{4.18}$$

Now

$$D = (a - 1) \sin bx - (a + 1) \cos bx \tag{4.19}$$

and we see that D determines not only the sign of M', but also the magnitude of M'. Moreover, $M' = 0$ when $D = 0$, and this happens when $(a - 1) \sin bx = (a + 1) \cos bx$, meaning

$$\tan bx = \frac{a + 1}{a - 1} = \frac{(1 + bx) + b^2}{1 + bx - b^2} \tag{4.20}$$

Even if $a = 1$, meaning $b = [x + (x^2 + 4)^{1/2}]/2$, we have $\cot bx = 0$, implying $bx = \pi/2 + n\pi$. Therefore, (4.20) is valid for all possible values of the entity a.

Note that the fractional function in b on the right-hand side of (4.20) is an increasing function of b. Of course, $\tan bx$ itself also is an increasing function of the variable b in the regions defined.

Note that $M' > 0$ when $D > 0$ and that $M' < 0$ when $D < 0$. Thus, M is an increasing function of b when $D > 0$ and vice versa.

4.1.2. Uniform Load on the Beam

Here the deflection is given as

$$Y = w \frac{2 - \exp(-bn) \cos bn - \exp(-bm) \cos bm}{2k} \tag{4.21}$$

where w is the uniform load intensity, and m and n are the beginning and end locations of the load along the beam. Again, we take Y as a function of b only, holding all the other entities as constants. We will use Y' to denote the first partial derivative of Y with respect to b for convenience and simplicity. Thus,

$$Y' = w \frac{A + B}{2k} \tag{4.22}$$

where

$$A = m \exp(-bm) [\cos bm + \sin bm] \tag{4.23}$$

$$B = n \exp(-bn) [\cos bn + \sin bn] \tag{4.24}$$

and k is treated as a constant even though k is involved in the definition of b. We will come back to this point later.

For now, let us note that a sufficient condition for $Y' > 0$ is the simultaneous satisfaction of the two conditions $A > 0$ and $B > 0$. This means

$$\cos bm + \sin bm > 0 \tag{4.25}$$

and

$$\cos bn + \sin bn > 0 \tag{4.26}$$

at the same time. Expressions (4.25) and (4.26) can be rewritten as

$$\tan bm > -1 \tag{4.25A}$$

$$\tan bn > -1 \tag{4.26A}$$

respectively.

Similarly, a sufficient condition for $Y' < 0$ is the simultaneous satisfaction of both $A < 0$ and $B < 0$, meaning

$$\tan bm < -1 \tag{4.27}$$

and

$$\tan bn < -1 \tag{4.28}$$

at the same time.

Of course, a sufficient condition for $Y' = 0$ is to have

$$A = B = 0 \tag{4.29}$$

leading to

$$\tan bm = \tan bn = -1 \tag{4.30}$$

The last formula may look odd, but it is a perfectly legitimate statement by noting the following fact: $\tan u = \tan v$ does not necessarily imply $u = v$. We know that bm is not equal to bn, but it is perfectly alright to have (4.30) in view of the point just made above.

It was mentioned above that we would come back to the discussion of treating k as a constant while using b as the parameter in our study. What this really means is that, in this context, b is a function of J only. Note that $db/dJ < 0$ from the definition of b given earlier in the last section. Thus,

$$[Y,J] > 0 \text{ if } [Y,b] < 0 \tag{4.31}$$

and conversely.

Also,

$$[Y,J] < 0 \text{ if } [Y,b] > 0 \tag{4.32}$$

and conversely.

Note that in the above and in the future, $[Y,u]$ is used to denote the first partial derivative of Y with respect to u for convenience. Also, we have been using ordinary derivative notations all along when we mention ahead of time that we are dealing with only one variable in that particular discourse.

Enough for the remarks so far. Let us look at things from another point of view. If we start with $b^4 = k/(4J)$, taking J as a constant and k as a function of b, then

$$[Y,k] > 0 \ \text{ if } \ [Y,b] > 0 \tag{4.33}$$

and conversely.

Also we have

$$[Y,k] < 0 \ \text{ if } \ [Y,b] < 0 \tag{4.34}$$

and conversely.

An important special case is when $m = n = mL/2$. Here

$$Y = w \frac{1 - \exp\left(-\dfrac{bL}{2}\right)\cos\left(\dfrac{bL}{2}\right)}{k} \tag{4.35}$$

Thus,

$$[Y,b] = \left(\frac{w}{k}\right)\left\{ L \exp\left(\frac{-bL}{2}\right)\left[\cos\left(\frac{bL}{2}\right) + \sin\left(\frac{bL}{2}\right)\right]\right\} \tag{4.36}$$

We see from (4.36) that $[Y,b] > 0$ when

$$\cos\left(\frac{bL}{2}\right) + \sin\left(\frac{bL}{2}\right) > 0 \tag{4.37}$$

or

$$\tan\left(\frac{bL}{2}\right) > -1 \tag{4.37A}$$

Similarly, we have $[Y,b] < 0$ when

$$\cos\left(\frac{bL}{2}\right) + \sin\left(\frac{bL}{2}\right) < 0 \tag{4.38}$$

or

$$\tan\left(\frac{bL}{2}\right) < -1 \tag{4.38A}$$

Naturally, we have $[Y,b] = 0$ when

$$\cos\left(\frac{bL}{2}\right) + \sin\left(\frac{bL}{2}\right) = 0 \tag{4.39}$$

That is, when

$$\tan\left(\frac{bL}{2}\right) = -1 \tag{4.39A}$$

Let us check Y''. After some mathematical operations, we obtain

$$Y'' = \frac{-wL^2 \exp\left(\frac{-bL}{2}\right) \sin\left(\frac{bL}{2}\right)}{k} \tag{4.40}$$

We evaluate $\sin(bL/2)$ at the $bL/2$ value that satisfies (4.39A) for $bL/2 = 3\pi/4 + n\pi$ to reach the result

$$\sin\left(\frac{bL}{2}\right) > 0 \tag{4.41}$$

and note $\exp(-bL/2) > 0$ along with w, L, and k being all positive to conclude from (4.40) that $Y'' < 0$. Therefore, Y has a maximum here. We know that there are many values of $bL/2$ that satisfy (4.39A). As a result, there are just as many values of b that yield the same maximum value of Y for a given L. The problem can be viewed from other perspectives with individual interpretations reached. For example, we can take

$$\left(\frac{bL}{2}\right) = \frac{7\pi}{4} + n\pi \tag{4.42}$$

and obtain

$$\sin\left(\frac{bL}{2}\right) < 0 \tag{4.43}$$

while still satisfying (4.39A), thereby reaching the conclusion that we have

$$Y'' > 0 \tag{4.44}$$

which means that Y has a minimum here.

Let us consider the possibility of Y being negative or zero. We note that $\exp(-bL/2) > 1$ except when $b = 0$, in which case $\exp(-bL/2) = 1$. Immediately we see that when $b = 0$, both $\exp(-bL/2) = 1$ and $\cos(bL/2) = 1$ hold. Substituting these values into the expression for Y, we obtain $Y = 0$. But in actual situations, $b = 0$ is not likely to happen. By observation of (4.35), we know that Y is nonnegative. Also, we can reach the same conclusion by applying the result from Y' to the domain of b that is of interest to us here.

4.2. BEAMS OF SEMI-INFINITE LENGTH

4.2.1. A Concentrated Force and Moment Acting at the End of the Beam

Adopting the notation $G = \exp(-bx)$, we have the deflection as

$$Y = \left(\frac{G}{2b^3L}\right)(P \cos bx + FbM_0) \tag{4.45}$$

where

$$F = \sin bx - \cos bx \tag{4.46}$$

and P and M_0 are the applied concentrated force and bending moment respectively. If we ask whether Y can be zero or negative, the answer is yes. We have

$$Y = 0$$

when

$$P \cos bx + FbM_0 = 0 \tag{4.47}$$

In other words, Y is zero when the ratio between M_0 and P is equal to $-(\cos bx)/(bF)$, for bF being nonzero if we want to view it that way. Here we use the ratio M_0/P and not its inverse. The reason is that we like the nice dimension of "length" provided to us through M_0/P. However, we can use P/M_0 if we want to (see below).

For the case where $bx = 0$ and b is nonzero, meaning $x = 0$, (4.47) becomes

$$P + FbM_0 = 0, \quad \text{with } F = -1 \tag{4.48}$$

This means that P and M_0 are proportional to each other, with $P/M_0 = b$.

For the trivial and impractical case where $x = 0$ and $b = 0$ simultaneously, we have, from (4.47),

$$P = 0 \tag{4.49}$$

Now, what can we say about $Y > 0$ and $Y < 0$? $Y > 0$ when

$$P \cos bx + FbM_0 > 0 \tag{4.50}$$

or

$$\frac{P}{M_0} > (1 - \tan bx)b \tag{4.51}$$

if we so desire, as long as we are careful about the many kinds of behaviors of $\cos bx$ and $\sin bx$, at different bx values, regarding positive, negative, or zero values and their implications when we come to do mathematical operations involving divisions and inequality. A naive interpretation of (4.50) is to take all the entities on the left-hand side of (4.50) as positive; then we surely automatically satisfy (4.50). When P

and M_0 are given as positive, then all it takes to do the trick is to keep both cos bx and F positive, and this is a sufficient condition for ensuring $Y > 0$.

As for $Y < 0$, the required condition is naturally

$$P \cos bx + bFM_0 < 0 \tag{4.52}$$

or

$$\frac{P}{M_0} < (1 - \tan bx)b \tag{4.53}$$

From (4.52), we see that a sufficient condition for $Y < 0$ is for both F and cos bx to be negative simultaneously, given that M_0 and P are both positive.

Furthermore, F is of critical importance, since first of all

$$F < 0 \text{ implies } 1 - \tan bx > 0 \tag{4.54}$$

$$F > 0 \text{ signifies } 1 - \tan bx < 0 \tag{4.55}$$

have far-reaching consequences as will be exemplified below. Second,

$$F = 0 \text{ means } \tan bx = 1 \tag{4.56}$$

which is a very special and important case to be explored in detail below.

Regarding $Y > 0$, the requirement is (4.51). If $F < 0$, then (4.51) may be satisfied under certain conditions. If $F > 0$, comparing (4.55) and (4.51), we see that (4.51) is unconditionally satisfied as long as $P/M_0 > 0$.

The requirement for $Y < 0$ is (4.53). We see that $F < 0$ is compatible with (4.53). When $F > 0$ is the case, then (4.53) may be satisfied if $P/M_0 < 0$.

Now let us look at the special case where $F = 0$. This means from (4.56) that

$$bx = \frac{\pi}{4} + n\pi \tag{4.57}$$

The consequences are threefold. First, Formulae (4.45), (4.47), (4.50), and (4.52) are all reduced to much simpler formulae, each of which consists of only one term, with the M_0 term missing. Second, Formula (4.50) is simplified to

$$P > 0 \tag{4.58}$$

which is free of M_0. Finally, Formula (4.52) is simplified to

$$P < 0 \tag{4.58A}$$

which is, again, independent of M_0.

Now let us look at an interesting special case, namely $x = 0$. Here

$$Y = \frac{P - bM_0}{2b^3 J} \tag{4.59}$$

Thus,

$$[Y,b] = \frac{2bM_0 - 3P}{2Jb^4} \tag{4.60}$$

from which we see that $[Y,b] < 0$ when

$$2bM_0 - 3P < 0 \tag{4.61}$$

and $[Y,b] > 0$ when

$$2bM_0 - 3P > 0 \tag{4.62}$$

Finally, in relation to the determination of extreme values of M, we find that $[Y,b] = 0$ when

$$2bM_0 - 3P = 0 \tag{4.63}$$

That is, when

$$b = \frac{3P}{2M_0} \tag{4.64}$$

for nonzero M_0.

Now we check the second derivative of Y with respect to b and find that it is positive if M_0 is positive and vice versa. Thus, for $M_0 > 0$, the value of b that satisfies (4.64) gives Y a minimum value, and of course, for $M_0 < 0$ we have a negative second derivative of Y with respect to b and hence a maximum value for Y.

Last but not least, we check $Y < 0$, $Y > 0$, and $Y = 0$ conditions. $Y = 0$ occurs when

$$P = bM_0 \tag{4.65}$$

$Y < 0$ is the case when

$$P < bM_0 \tag{4.66}$$

$Y > 0$ holds when

$$P > bM_0 \tag{4.67}$$

It is interesting to note that $b = P/M_0$ in (4.65), but $b = 3P/(2M_0)$ in (4.64).

4.2.2. Uniform Load on the Beam with a Simply Supported End

The reaction is

$$R = \frac{w}{2b} \tag{4.68}$$

which is inversely proportional to b.

Looking into the elements in b, we see that R is inversely proportional to the fourth root of k and directly proportional to the fourth root of J.

The deflection is

$$y = \left(\frac{w}{k}\right)[1 - \exp(-bx) \cos bx] \tag{4.69}$$

This expression is of exactly the same mathematical form as in a special case, specified by $m = n = L/2$, of the problem for a beam of infinite length under uniform load. Therefore, similar conclusions are expected. Thus, we have

$$[Y,b] = \left(\frac{wGx}{k}\right)[\cos bx + \sin bx] \tag{4.70}$$

from which we have $[Y,b] < 0$ when

$$\cos bx + \sin bx < 0 \tag{4.71}$$

and $[Y,b] > 0$ when

$$\cos bx + \sin bx > 0 \tag{4.72}$$

Finally, $[Y,b] = 0$ when

$$\cos bx + \sin bx = 0 \tag{4.73}$$

or

$$\tan bx = -1 \tag{4.74}$$

The second derivative of Y with respect to b evaluated at the bx that satisfies (4.74) is negative, so Y has a maximum here.

When will we have $Y = 0$? This happens when

$$\exp(-bx) \cos bx = 1 \tag{4.75}$$

There are two cases to consider.

Case 1. x = 0

As a result of this defining condition, one possibility is $\cos bx = 1$, $\exp(-bx) = 1$ simultaneously. Thus, $bx = 0, 2\pi\dots$. The other possibility is $\exp(-bx) = 1/\cos bx$.

Case 2. x = Nonzero

We have $\exp(-bx) \neq 1$, $\exp(bx) > 1$, and $\exp(-bx) < 1$. We need $\cos bx > 1$ to satisfy (4.75). But this is not possible because of the property of cosine function. Therefore, this case does not exist. In other words, this case does not have physical meaning.

Next, when will we have $Y > 0$? This happens when

$$1 > \exp(-bx) \cos bx \qquad (4.76)$$

meaning

$$\frac{1}{G} > C, \quad \text{with } C = \cos bx, \; G = \exp(-bx) \qquad (4.77)$$

But

$$G > 1 \qquad (4.78)$$

in general and

$$G = 1 \qquad (4.79)$$

only when $bx = 0$, which is excluded from (4.76). Thus, (4.76) or (4.77) is established, meaning $C < 1$, as the defining condition for $Y > 0$.

What about $Y < 0$? This requires

$$1 < GC \qquad (4.80)$$

meaning

$$\frac{1}{G} < C \qquad (4.81)$$

But this is not possible, for two reasons. First, we have $1/G > 1$, except $bx = 0$, in which case we have $Y = 0$ already and is, therefore, excluded from our consideration here. Second, we know that C can never be greater than 1. Then how can C be greater than $1/G$? Therefore, (4.80) is not satisfied, and it is not possible to have $Y < 0$.

4.2.3. Uniform Load on the Beam with a Fixed End

It is interesting to note that the reaction here is twice that in the case of a simply supported end. That is,

$$R = \frac{w}{B} \qquad (4.82)$$

Naturally, it has all the characteristics described in the last section about R.

Let us look at the bending moment

$$M = \frac{-2b^2 Jw}{k} \qquad (4.83)$$

which is negative.

Since b, k, and J are related via

$$b = \left[\frac{k}{4J}\right]^{1/4} \tag{4.84}$$

we may rewrite (4.83) as

$$M = -w\left(\frac{J}{k}\right)^{1/2} \tag{4.85}$$

Thus, we observe from (4.85) that in addition to being directly proportional to the uniform load intensity w, the bending moment M is directly proportional to the square root of J and inversely proportional to the square root of k.

We see also from the definition of b in terms of J, k that there are many ways to go about varying b, depending on our goals and plans. For example, we may keep J constant and vary k systematically, do the opposite, or even change both J and k simultaneously in a specific manner of choice. This can be done numerically using a computer or can be pursued further analytically to suit individual needs. This is just an example. There are many other problems that can be approached in this way, and the procedure outlines and methodology have been indicated previously.

Let us come back to the study of M. For keeping k as a constant, we have

$$M = \frac{-w}{2b^2} \tag{4.86}$$

which is inversely proportional to the square of b. For letting k vary with b via

$$k = 4Jb^2 \tag{4.87}$$

we have also the expression for M as shown in (4.86).

For our study of beams on elastic foundations, most of the subjects are considered as functions of one independent variable b, while all other entities are viewed as constants, including the longitudinal axis x. However, we know from the definition of b that b is dependent on two other entities, J and k. This point was touched upon above and will be elaborated on a little more now.

In order to assess the effects of the independent variables J and k on a subject under study, we may, first of all, obtain the results from considering the subject as a function of b only, as we did for most of the topics earlier. Then we may consider b as function of two variables J and k and obtain first and second partial derivatives of b with respect to J and k separately. Finally, we use the principles of calculus, including the chain rule, to obtain all the necessary partial derivatives of the subject function with respect to J and k, including the so-called mixed second partial derivatives, and evaluate the possible maximum and minimum values of subject functions using the information thus collected.

Based on the remarks just presented, the following is provided. Considering b as a function of two independent variables, J and k, we have the first partial derivative of b with respect to J as

$$[b,J] = \left(-\frac{1}{4}\right)\left(\frac{k}{4}\right)^{1/4} J^{-5/4} \qquad (4.88)$$

which is always negative (never zero or positive), signifying b as a decreasing function of J.

The second partial derivative of b with respect to J is

$$[b,JJ] = \left(\frac{5}{16}\right)\left(\frac{k}{4}\right)^{1/4} J^{-9/4} \qquad (4.89)$$

which is positive.

Now, we have

$$[b,k] = A(k)^{-3/4} \qquad (4.90)$$

where

$$A = \frac{1}{4(4J)^{1/4}} \qquad (4.91)$$

Note that $[b,k]$ is positive, and its value decreases as J increases and/or k increases, although at different rates, with the latter being the higher one.

Next we have

$$[b,kk] = \left(\frac{-3}{4}\right) A(k)^{-7/4} \qquad (4.92)$$

which is negative, and its "absolute" value decreases as either J or k increases, with k being the stronger player again.

Finally, a so-called mixed second partial derivative of b is

$$[b,Jk] = \left(-\frac{1}{4}\right)(4J)^{-5/4} k^{-3/4} \qquad (4.93)$$

which is negative also.

From (4.90) and (4.93), it is interesting to see that

$$[b,Jk] = -\frac{[b,k]}{4J} \qquad (4.94)$$

Of course, we can also rewrite (4.93) as

$$[b,Jk] = \left(\frac{-1}{64}\right)(J)^{-5/4}\left(\frac{k}{4}\right)^{-3/4} \tag{4.95}$$

Thus, we have another interesting result:

$$[b,Jk] = \frac{[b,J]}{4k} \tag{4.96}$$

Note that here we have

$$[b,Jk] = [b,kJ] \tag{4.97}$$

5

CANTILEVERS

5.1. UNIFORM LOAD ON PART OF THE SPAN

For the general case under the theme of this section, with nonzero a and c as shown in Figure 5.1, the results are as follows. The reaction is

$$R = wb \tag{5.1}$$

The bending moments are

$$M_1 = Rx - M_0 \quad \text{for } 0 < x < a \tag{5.2}$$

$$M_2 = Rx - M_0 - w\frac{(x-a)^2}{2} \quad \text{for } a < x < d \tag{5.3}$$

$$M_3 = 0 \quad \text{for } d < x < L \tag{5.4}$$

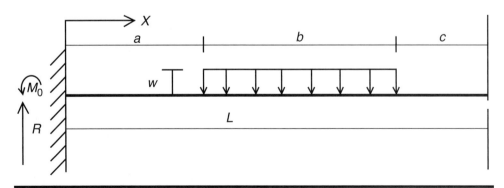

FIGURE 5.1. Uniform load on a cantilever

where

$$d = a + b \tag{5.5}$$

The moment at the fixed end is

$$-M_0 = -wb\left(a + \frac{b}{2}\right) \tag{5.6}$$

The deflections are

$$Y = \left[\frac{1}{EI}\right]\left[\frac{-M_0 x^2}{2} + \frac{Rx^3}{6}\right] \quad \text{for } 0 < x < a \tag{5.7}$$

$$Y = \left[\frac{1}{EI}\right]\left[\frac{-M_0 a^2}{2} + \frac{Ra^3}{6}\right] \quad \text{at } x = a \tag{5.8}$$

$$Y = \left[\frac{1}{EI}\right]\left\{A + B + (x - a)^2\left[R\frac{x + 2a}{6} - \frac{M_0}{2} - w\frac{(x - a)^2}{24}\right]\right\} \tag{5.9}$$

for $a < x < d$

where

$$A = -M_0 a\left(x - \frac{a}{2}\right), \qquad B = Ra^2\left[\frac{\left(x - \frac{2b}{3}\right)}{2}\right] \tag{5.10}$$

$$Y = \left[\frac{1}{EI}\right]\left\{C + D + b^2\left[R\frac{3a + b}{6} - \frac{M_0}{2} - \frac{wb^2}{24}\right]\right\} \quad \text{at } x = d \tag{5.11}$$

where

$$C = M_0 a\left(\frac{a}{2} + b\right), \qquad D = Ra^2\left(\frac{\frac{a}{3} + b}{2}\right) \tag{5.12}$$

$$Y = \left[\frac{1}{EI}\right][Fx + G] \quad \text{for } d < x < L \tag{5.13}$$

$$Y = \left[\frac{1}{EI}\right][FL + G] \quad \text{at } x = L \tag{5.14}$$

where

$$F = \frac{Ra^2}{2} - M_0 a - \frac{wb^3}{6} \tag{5.15}$$

$$G = -\frac{Ra^3}{3} + \frac{M_0 a^2}{2} + wb^3(4a + b) \tag{5.16}$$

Some important special cases are as follows.

Case 1. a = 0

First of all, note that $A = B = C = D = 0$ because $a = 0$. The reaction is

$$R = wb \tag{5.17}$$

The bending moments are

$$M_1 = Rx - M_0 \quad \text{for } 0 < x < b \tag{5.18}$$

$$M_2 = Rx - M_0 - \frac{wx^2}{2} \quad \text{for } b < x < L \tag{5.19}$$

The moment at the fixed end is

$$-M_0 = \frac{-wb^2}{2} \tag{5.20}$$

The deflections are given by (5.9–5.16) with simplified results and

$$d = b \tag{5.21}$$

$$F = \frac{-wb^3}{6} \tag{5.22}$$

$$G = wb^4 \tag{5.23}$$

Thus, the deflections are

$$Y = \left[\frac{1}{EI}\right]\left[\frac{-M_0}{2} + \frac{Rx}{6} - \frac{wx^2}{24}\right]x^2 \quad \text{for } 0 < x < b \tag{5.24}$$

$$Y = \left[\frac{1}{EI}\right]\left[\frac{-M_0}{2} + \frac{Rb}{6} - \frac{wb^2}{24}\right]b^2 \quad \text{at } x = b \tag{5.25}$$

$$Y = \left[\frac{1}{EI}\right][Fx + G] \quad \text{for } b < x < L \tag{5.26}$$

$$Y = \left[\frac{1}{EI}\right][FL + G] \quad \text{at } x = L \tag{5.27}$$

with F and G given by (5.22) and (5.23) respectively.

Case 2. c = 0

The reaction is

$$R = wb \tag{5.28}$$

The bending moments are

$$M_1 = Rx - M_0 \quad \text{for } 0 < x < a \tag{5.29}$$

$$M_2 = Rx - M_0 - w\frac{(x - a)^2}{2} \quad \text{for } a < x < L \tag{5.30}$$

The moment at the fixed end is

$$-M_0 = -wb\left(a + \frac{b}{2}\right) \tag{5.31}$$

The deflections are given by (5.7–5.12) with $d = L$.

Case 3. a = 0, c = 0, and b = L = d

The reaction is

$$R = wL \tag{5.32}$$

The bending moments are

$$M = Rx - M_0 \quad \text{for } 0 < x < L \tag{5.33}$$

The moment at the fixed end is

$$-M_0 = \frac{-wL^2}{2} \tag{5.34}$$

The deflections are

$$Y = \left[\frac{1}{EI}\right]\left[\frac{-M_0}{2} + \frac{Rx}{6} - \frac{wx^2}{24}\right]x^2 \quad \text{for } 0 < x < L \tag{5.35}$$

with

$$\max Y = \left[\frac{1}{EI}\right]\left[-\frac{wL^4}{8}\right] \quad \text{at } x = L \tag{5.36}$$

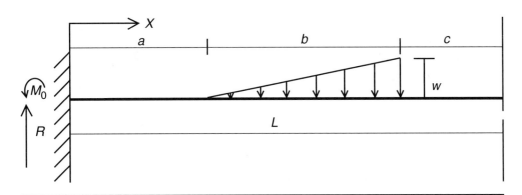

FIGURE 5.2. Triangular load on a cantilever

5.2. TRIANGULAR LOAD ON PART OF THE SPAN

For the general case with nonzero a and c as shown in Figure 5.2, the results are as follows. The reaction is

$$R = \frac{wb}{2} \tag{5.37}$$

The bending moments are

$$M_1 = Rx - M_0 \quad \text{for } 0 < x < a \tag{5.38}$$

$$M_2 = Rx - M_0 - w\frac{(x-a)^3}{6b} \quad \text{for } a < x < d \tag{5.39}$$

$$M_3 = 0 \quad \text{for } d < x < L \tag{5.40}$$

where

$$d = a + b \tag{5.41}$$

The moment at the fixed end is

$$-M_0 = -R\left(a + \frac{2b}{3}\right) \tag{5.42}$$

The deflections are

$$Y = \left[\frac{1}{EI}\right]\left[\frac{-M_0 x^2}{2} + \frac{Rx^3}{6}\right] \quad \text{for } 0 < x < a \tag{5.43}$$

$$Y = \left[\frac{1}{EI}\right]\left[\frac{-M_0 a^2}{2} + \frac{Ra^3}{6}\right] \quad \text{at } x = a \tag{5.44}$$

$$Y = \left[\frac{1}{EI}\right]\left[Rd^2\,\frac{3x - 2d}{6} - \frac{M_0 x^2}{2} - w\,\frac{(x - a)^5}{120b}\right] \tag{5.45}$$

for $a < x < d$

$$Y = \left[\frac{1}{EI}\right]\left[\frac{Rd^3}{6} - \frac{M_0 d^2}{2} - \frac{wb^4}{120}\right] \quad \text{at } x = d \tag{5.46}$$

$$Y = \left[\frac{1}{EI}\right]\left\{F - wb^3\,\frac{5(x - a) - 4b}{120}\right\} \quad \text{for } d < x < L \tag{5.47}$$

$$Y = \left[\frac{1}{EI}\right][G + H] \quad \text{at } x = L \tag{5.48}$$

where

$$F = \left(x - \frac{2a}{3}\right)\frac{Ra^2}{2} - M_0 a\left(x - \frac{a}{2}\right) \tag{5.49}$$

$$G = wab\,\frac{a^2 + 2ab - L(3a + 4b)}{12} \tag{5.50}$$

$$H = -wb^3\,\frac{5(L - a) - 4b}{120} \tag{5.51}$$

Some important special cases are as follows.

Case 1. a = 0 and d = b

The reaction is given by (5.37). The bending moments are

$$M_1 = Rx - M_0 - \frac{wx^3}{6b} \quad \text{for } 0 < x < b \tag{5.52}$$

$$M_2 = 0 \quad \text{for } b < x < L \tag{5.53}$$

The moment at the fixed end is

$$-M_0 = \frac{-2Rb}{3} \tag{5.54}$$

The deflections are

$$Y = \left[\frac{1}{EI}\right]\left[Rb^2\,\frac{3x - 2b}{6} - \frac{M_0 x^2}{2} - \frac{wx^5}{120b}\right] \quad \text{for } 0 < x < b \tag{5.55}$$

$$Y = \left[\frac{1}{EI}\right]\left[\frac{Rb^3}{6} - \frac{M_0 b^2}{2} - \frac{wb^4}{120}\right] \quad \text{at } x = b \tag{5.56}$$

$$Y = \left[\frac{1}{EI}\right]\left[-wb^3 \frac{5x - 4b}{120}\right] \quad \text{for } b < x < L \tag{5.57}$$

$$Y = \left[\frac{1}{EI}\right]\left[-wb^3 \frac{5L - 4b}{120}\right] \quad \text{at } x = L \tag{5.58}$$

Case 2. c = 0

By definition, we have $a + b = d = L$. The reaction is given by (5.37). The bending moments are given by (5.38) and (5.39). Note that (5.40) does not appear here. The moment at the fixed end is given by (5.42).

The deflections are given by (5.43–5.46) with $a + b = d = L$. Specifically:

$$Y = \left[\frac{1}{EI}\right]\left[RL^2 \frac{3x - 2L}{6} - \frac{M_0 x^2}{2} - w \frac{(x - a)^5}{120b}\right] \tag{5.59}$$

$$\text{for } a < x < L$$

$$Y = \left[\frac{1}{EI}\right]\left[\frac{RL^3}{6} - \frac{M_0 L^2}{2} - \frac{wb^4}{120}\right] \quad \text{at } x = L \tag{5.60}$$

Case 3. a = 0, c = 0, and d = b = L

The reaction is given by (5.37). The bending moments are

$$M = Rx - M_0 \frac{wx^3}{6b} \quad \text{for } 0 < x < L \tag{5.61}$$

The moment at the fixed end is

$$-M_0 = \frac{-2RL}{3} \tag{5.62}$$

The deflections are

$$Y = \left[\frac{1}{EI}\right]\left[RL^2 \frac{3x - 2L}{6} - \frac{M_0 x^2}{2} - \frac{wx^5}{120L}\right] \quad \text{for } 0 < x < L \tag{5.63}$$

with

$$\text{max } Y = \left[\frac{1}{EI}\right]\left[-\frac{M_0}{2} + \frac{RL}{6} - \frac{wL^2}{120}\right]L^2 \quad \text{at } x = L \tag{5.64}$$

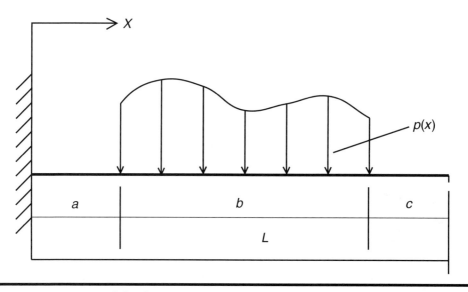

FIGURE 5.3. General load on a cantilever (for general theory)

5.3. GENERAL LOAD INTENSITY FUNCTIONS WITH APPLICATIONS

Consider an arbitrary load intensity function denoted by $p(x)$. The beam under such a load is depicted in Figure 5.3.

The reaction is

$$R = \int_a^e p(x)dx \qquad (5.65)$$

The end moment is

$$M_0 = -\int_a^e xp(x)dx \qquad (5.66)$$

where

$$e = a + b \qquad (5.66A)$$

The bending moments are

$$M_1 = Rx + M_0 \quad \text{for } 0 < x < a \qquad (5.67)$$

$$M_2 = Rx - B + M_0 \quad \text{for } a < x < e \qquad (5.68)$$

$$M_3 = 0 \quad \text{for } x > e \qquad (5.69)$$

where

$$B = \int_a^x (x - t)p(t)dt \tag{5.70}$$

For convenience of operations in the following development regarding deflections, let us define

$$C = \int_0^a M_1(t)dt + \int_a^e M_2(t)dt \tag{5.71}$$

$$D = \int_0^a (L - t)M_1(t)dt + \int_a^e (L - t)M_2(t)dt \tag{5.72}$$

$$J = EI \tag{5.73}$$

The deflection Y for each region can be obtained from the following. The deflection is considered to be positive when it is downward. For the region $0 < x < a$

$$JY = \int_0^x (x - t)M_1(t)dt \tag{5.74}$$

For the region $a < x < e$

$$JY = \int_0^a (x - t)M_1(t)dt + \int_a^x (x - t)M_2(t)dt \tag{5.75}$$

For the region $e < x < L$

$$JY = C(x - c) \tag{5.76}$$

where

$$c = L - \frac{D}{C} \tag{5.77}$$

Note that D defined by Expression (5.72) is the deflection at the tip of the cantilever with a multiplier J. For the purpose of determining the entity c needed in (5.76), we could use other approaches. For example, we could define G, an expression similar to D, such that $c = G/C$. However, then the entity G so defined does not have the physical significance revealed by D, even though G has a little bit simpler look and may lead to a simpler expression for the distance c.

Some special cases of interest are as follows.

Case 1. a = 0

The immediate consequences are that $e = b$, M_1 disappears, and the entities C and D are simplified as

$$C = \int_0^b M_2(t)dt \tag{5.78}$$

$$D = \int_0^b (L - t)M_2(t)dt \tag{5.79}$$

where M_2 is still given by Formula (5.68) but with

$$R = \int_0^b p(x)dx \tag{5.80}$$

$$M_0 = -\int_0^b xp(x)dx \tag{5.81}$$

$$B = \int_0^x (x - t)p(t)dt \tag{5.82}$$

The deflections Y for the two regions can be obtained in the following manner. For the region $0 < x < b$

$$JY = \int_0^x (x - t)M_2(t)dt \tag{5.83}$$

For the region $b < x < L$, the formal formula for the general case still applies but with a simplified C and c with which to work.

Case 2. a + b = L and a ≠ 0

We have, as a result of the defining condition, first of all, $e = L$, then M_1 and M_2 are as before in the general case formally but with the region corresponding to M_3 disappearing. Therefore, we have the reaction

$$R = \int_0^L p(x)dx \tag{5.84}$$

while the bending moments as well as the deflections formulae for the first two regions are as in the general case except that the entity e is replaced by the span length L. Additionally, we have

$$M_0 = -\int_0^L xp(x)dx \qquad (5.85)$$

The third region defined by $e < x < L$ drops out and so do Formulae (5.76) and (5.77) for this particular case. Here, D is as given by Formula (5.72) in the general case but with the entity e replaced by the span length L.

Case 3. *a = 0 and b = L*

This is a further simplification from case 2 by setting $a = 0$ in all the formulae in that case. As a result of this simplification, we have the following interesting situation. The bending moment M_1 disappears along with the first region, and so does the third region, whereas the second region occupies the entire span. In other words, this important case corresponds to a cantilever with the entire span being loaded. Thus, we have the reaction

$$R = \int_0^L p(x)dx \qquad (5.86)$$

and the end moment

$$M_0 = -\int_0^L xp(x)dx \qquad (5.87)$$

The entity B is still formally given by Formula (5.82), whereas

$$C = \int_0^L M_2(t)dt \qquad (5.88)$$

with M_2 given formally by Formula (5.68).

Finally, we have the following two important formulae

$$D = \int_0^L (L - t)M_2(t)dt \qquad (5.89)$$

and

$$JY = \int_0^x (x - t)M_2(t)dt \qquad (5.90)$$

The former signifies the deflection at the tip of the cantilever with a multiplier J, while the latter represents the general expression for deflection, with the same multiplier J.

A remark on the approach taken here is in order. Another way to solve the same problem is by resorting to utilization of the results of a single concentrated force and subsequent integration.

Illustrative Examples

Let us change pace in order to arouse the reader's interest by taking a slightly different approach. The load intensity function will be defined by a local coordinate system t as shown in Figure 5.4, which is presented mainly for symbols and notations and not for the actual shape of the load intensity function. The following are illustrative examples presented with details.

Example 1

The load intensity function is a general quadratic polynomial of the form

$$p(t) = at^2 + bt + c \qquad (5.91)$$

where a, b, and c are constants independent of t.

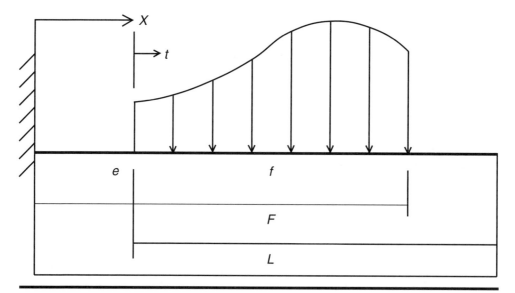

FIGURE 5.4. General load on a cantilever (for the illustrative examples)

We note immediately that the solution to this problem can also be obtained from superposition of the two problems specified below. One is the problem associated with the general linear load intensity function presented previously. The other is the one corresponding to the load intensity function $q(t) = gt^2$, where g is a constant independent of t. However, we will stick to the load intensity function represented in (5.91) in order to examine another view of the problem and its solution. The results from both of these approaches are, of course, exactly the same.

The reaction is

$$R = \left(\frac{a}{3}\right)f^3 + \left(\frac{b}{2}\right)f^2 + cf \tag{5.92}$$

and the end moment is

$$M_0 = eR + \left(\frac{a}{4}\right)f^4 + \left(\frac{b}{3}\right)f^3 + \left(\frac{c}{2}\right)f^2 \tag{5.93}$$

We see that both R and M_0 are increasing functions of the variable f, the extent of the loaded region, when the constants a, b, and c are all positive as a sufficient condition and when e is fixed. Also, it is very interesting to observe the following. When we put

$$M_0 = RX \tag{5.94}$$

where

$$X = e + nf \tag{5.95}$$

with

$$n = \frac{\left(\frac{a}{4}\right)f^3 + \left(\frac{b}{3}\right)f^2 + \left(\frac{c}{2}\right)f}{\left(\frac{a}{3}\right)f^2 + \left(\frac{b}{2}\right)f + c} \tag{5.95A}$$

then we have the first derivative of M_0 with respect to f as

$$M_0' = (e + f)R' \tag{5.96}$$

where

$$R' = af^2 + bf + c \tag{5.97}$$

which is exactly the value of the load intensity function at $t = f$; that is,

$$R' = p(f) \tag{5.98}$$

From (5.96), M_0' is directly proportional to both R' and the distance $(e + f)$ and is always positive for positive a, b, and c. The last statement confirms the observation made earlier about M_0. The case where some of the constants a, b, and c are negative can be explored in a similar fashion as follows.

Since R is of fundamental importance and many other entities, for example bending moments, depend on R directly, it is expedient to deal with the investigation of R first. Thus, we will proceed in the following manner.

There exist conditions under which R' may become zero or negative when some of the constants a, b, and c are not positive. Let us examine the situation in which $R' = 0$. This happens when

$$af^2 + bf + c = 0 \tag{5.99}$$

We conclude immediately that

$$f = \frac{-b \pm (b^2 - 4ac)^{1/2}}{2a} \tag{5.100}$$

For definiteness and consistency in the following development, let us define

$$Q = (b^2 - 4ac)^{1/2} \tag{5.100A}$$

and call the two roots in (5.100) separately as

$$f_1 = \frac{-b + Q}{2a} \tag{5.100B}$$

$$f_2 = \frac{-b - Q}{2a} \tag{5.100C}$$

In order for f_1 or f_2 to have the positive real value required by the physical considerations of the problem, we have, first of all,

$$b^2 - 4ac > 0 \tag{5.101}$$

or

$$b^2 - 4ac = 0 \tag{5.102}$$

Let us focus on the case defined by Expression (5.101) first and consider the most general situation, namely where c is nonzero. For the sake of clarity and being systematic, we will proceed with our investigation category by category as follows.

Category A. a > 0

This means that the coefficient of the highest term in the expression for R' is positive. Here we need to consider four cases separately to gain insight into the defining conditions for revealing the underlying physical significance.

Case 1. a > 0 and b > 0. The defining property of this case dictates that Equation (5.99) can have only one positive real root, and this occurs when $Q > b$.

This requires that $4ac < 0$. However, since $a > 0$ is given, we must have $c < 0$. We have one such root, f_1.

Case 2. $a > 0$ and $b = 0$. Once again, in view of Q and the requirement of a real and positive root, it is necessary that $4ac < 0$. Therefore, a and c must have opposite signs. Since it is given that $a > 0$, it follows that $c < 0$. In this case, we also have only one positive real root, f_1.

Let us consider the situation where $a > 0$ and $b < 0$ by trying to follow suit from above. It is interesting to discover that the task is not as simple as the two cases just encountered. Rather, we have the following. Depending on whether c is negative or positive, we have two different cases. Note that the case $c = 0$ will be dealt with separately later.

Case 3. $a > 0$, $b < 0$, and $c < 0$. As before, only one root, f_1, is admissible, because $Q > -b$ is required.

Case 4. $a > 0$, $b < 0$, and $c > 0$. This implies that $Q < -b$. Note that $-b > 0$ and we have two positive real roots, f_1 and f_2, available.

Category B. $a < 0$

Here, as in category A above, we will consider four cases

Case 5. $a < 0$, $b > 0$, and $c < 0$. We have $Q < b$. This is because $a < 0$ and $c < 0$. Thus, $ac > 0$ and $b^2 > b^2 - 4ac$. By examining the expressions for f_1 and f_2, we see that we can have two positive real roots.

Case 6. $a < 0$, $b > 0$, and $c > 0$. In this case, we have $Q > b$ due to the fact that $ac < 0$. This leads to $-b + Q > 0$, which, together with the given condition that $a < 0$, will render $-b + Q/(2a)$, a negative value for x. However, this is not what we want. Therefore, only one root, f_2, is admissible.

Case 7. $a < 0$, $b = 0$, and $c > 0$. For Q to have a real value, it is necessary that $ac < 0$. However, $a < 0$ is given, so we must have $c > 0$, and this leads to just one root, f_2. The other root is not admissible based on considerations of physical meaning of the roots of the equation to be solved.

Case 8. $a < 0$, $b < 0$, and $c > 0$. This corresponds to the condition where $Q > -b > 0$ and only one root, f_2, is admissible. Note that we might play with the case $a < 0$, $b < 0$, and $c < 0$ and see what is happening. We may be surprised to find that there is not even one root that is admissible.

Thus far, we have examined the most general category which corresponds to a nonzero c. There are eight cases in this category. Each of these cases provides us with information about a critical point or two for a specified range of values of the coefficients a, b, and c of the load intensity function p or even the reaction R. For cases 4 and 5, there are two critical points for each. The details of these critical points and their significance will be given below. This can best be achieved by considering categories A and B while utilizing the convention adopted for the designation of the two roots of equation $R' = 0$ in (5.99).

For category A, defined by $a > 0$, we have $f_1 > f_2$ to signify the relative magnitude of the two roots of the equation under consideration. Here, R has a minimum value at $f = f_1$ for all the possible values of b (positive, negative, and zero) and nonzero c. Moreover, R has the only maximum value at $f = f_2$ when the conditions $b < 0$ and $c > 0$ are satisfied simultaneously.

For category B, defined by $a < 0$, we have $f_1 < f_2$, which shows the relative magnitude of the two roots of $R' = 0$. In this group, R has a maximum at $f = f_2$ for all the values of b (positive, negative, and zero) and nonzero c. Moreover, R has the only minimum value at $f = f_1$ when both the conditions $b > 0$ and $c < 0$ are satisfied simultaneously.

Now let us take a look at the special case where $c = 0$. Here, the equation to be solved is simplified to

$$af^2 + bf = 0 \tag{5.103}$$

This equation has two roots:

$$f_1 = 0 \tag{5.103A}$$

and

$$f_2 = -\frac{b}{a} \tag{5.103B}$$

However, the first root does not have any physical meaning. The second one is physically meaningful only when both conditions $b \neq 0$ and $-b/a > 0$ are satisfied. This means that the coefficients a and b must have opposite signs. The implication is that we actually have two cases here, namely case 9 and case 10.

Case 9. c = 0, a < 0, and b > 0. R has a maximum value at $f = -b/a$.

Case 10. c = 0, a > 0, and b < 0. R has a minimum value at $f = -b/a$.

If, additionally, $a = 0$ or $b = 0$, then we have $f = 0$, which is a trivial load case.

Thus far, we have examined all the cases in the most general category of nonzero c under the premise that Q is nonzero. What happens when Q is zero? This question can be answered in more than one way. We will answer it in a rather unusual way.

Consider the second derivative of R with respect to f and set it to zero. We obtain

$$2af + b = 0 \tag{5.104}$$

This will lead to the solution

$$f = -\frac{b}{2a} \tag{5.104A}$$

Now look at the solution to $R' = 0$ presented in (5.100). We see that Formula (5.104A) corresponds to the situation where $Q = 0$ in (5.100), but this means that, from (5.100),

$$b^2 = 4ac \tag{5.105}$$

We see that the solution presented in (5.104A) is for both $R' = 0$ and $R'' = 0$ and therefore is a point of inflection for the function R. Moreover, since both a and c are nonzero, b^2 is also nonzero and positive by virtue of (5.105). Bearing this in mind and looking at (5.105) again, we observe that there are two cases in this category specified by (5.104). They are as follows.

Case 11. $a < 0$, $b > 0$, and $c < 0$. R is a decreasing function of the independent variable f except at the point of inflection.

Case 12. $a > 0$, $b < 0$, and $c > 0$. R is an increasing function of f, except at the point of inflection.

It is interesting to note here that whether the highest coefficient in R is positive or negative actually determines whether R is an increasing or decreasing function of f respectively. This concludes the investigation of the reaction R.

The bending moments are

$$M = R(X - x) \quad \text{for } x < e \tag{5.106}$$

When $x = 0$, we have

$$M = M_0 = RX \tag{5.106A}$$

When $x = e$, we have

$$M = R(nf) \tag{5.106B}$$

For the region $e < x < e + f$

$$M = \left(\frac{a}{4}\right)g_4 + \left(\frac{b - ay}{3}\right)g_3 + \left(\frac{c - by}{2}\right)g_2 - cyg_1 \tag{5.107}$$

where

$$g_i = f^i - t^i, \quad i = 1, 2, 3 \tag{5.108}$$

and t is the local horizontal axis defined in Figure 5.5 (see Section 5.4).

When $x = e$, $t = 0$, we have from (5.108)

$$g_i = f^i \tag{5.108A}$$

with the simplified $M = R(nf)$, which confirms the result shown in (5.106B), as it should.

When $t = f$, $x = e + f$, we have $g_i = 0$, so

$$M = 0 \tag{5.109}$$

For the region $x > e + f$

$$M = 0 \tag{5.110}$$

The deflections Y can be obtained from the following. The deflection at the tip of the cantilever is

$$Y = Y_1 + Y_2 \tag{5.111}$$

where

$$Y_1 = \left(\frac{R}{J}\right) \sum_{i=1}^{3} (E_i) e^i \tag{5.112}$$

with

$$E_1 = XL, \qquad E_2 = -\frac{L + X}{3}, \qquad E_3 = \frac{1}{3} \tag{5.113}$$

and the E_i's have absolutely nothing to do the E_i's in Chapters 3 and 6 and the appendices.

$$Y_2 = \left(\frac{1}{J}\right) \sum_{i=1}^{6} (F_i) f^i \tag{5.114}$$

with

$$F_1 = nfRF, \qquad F_2 = -(nf + F)\frac{R}{2},$$

$$F_3 = \frac{\dfrac{cF}{2} + R}{3}, \qquad F_4 = \frac{\dfrac{bF}{6} - \dfrac{c}{2}}{4} \tag{5.115}$$

$$F_5 = \frac{\dfrac{aF}{12} - \dfrac{b}{6}}{5}, \qquad F_6 = -\frac{a}{72} \tag{5.116}$$

Some special cases of interest and practical importance are as follows.

Case 1. e = 0 with General f

R remains the same as in the general case. M_0 simplifies to

$$M_0 = nfR = m = \frac{af^4}{4} + \frac{bf^3}{3} + \frac{cf^2}{2} \tag{5.117}$$

with

$$X = \frac{m}{R} \tag{5.118}$$

and region $x < e$ disappearing.

Region 0 < x < f. M remains the same as in the general case. At $x = 0$, M = M_0, as it should be. At $x = f$, $M = 0$ as in the general case.

Region x > f. $M = 0$ as in the general case.

Case 2. f = F

This means that the load is extended to the tip of the cantilever. Hence, R is given formally as in the general case, with f replaced by F. M_0 is also formally given as in the general case, with f replaced by F. Specifically, we have the following.

Region x < e. Formal expressions for R and M remain the same as in the general case, with all the f's in R,X replaced by F.

Region e < x < e + f. This becomes $e < x < L$, because here $e + F = L$ and $f = F$. M is the same as in the general case, except f is replaced by F.

At $x = e$, $t = 0$, and therefore, we have

$$M = \left(\frac{a}{4}\right)F^4 + \left(\frac{b}{3}\right)F^3 + \left(\frac{c}{2}\right)F^2 \tag{5.119}$$

As for the deflection at the tip, we can obtain it from the general case by replacing all the f's with F.

Case 3. e = 0 and f = F

This means that $f = L = F$. Therefore, we have

$$R = \left(\frac{a}{3}\right)L^3 + \left(\frac{b}{2}\right)L^2 + (c)L \tag{5.120}$$

$$M_0 = \left(\frac{a}{4}\right)L^4 + \left(\frac{b}{3}\right)L^3 + \left(\frac{c}{2}\right)L^2 \tag{5.121}$$

M at any section is the same as in the general case, except f is replaced by L. Thus, at $x = 0$, $M = M_0$, as shown in (5.121). At $x = L$, $M = 0$ as before.

As for the deflection at the tip of the cantilever, part $Y_1 = 0$, because $e = 0$; part Y_2 is obtained from the general case, with f replaced by L.

Example 2

The load intensity function is a third-degree polynomial of the form

$$q = a + bt + ct^2 + gt^3 \tag{5.122}$$

Once again, in view of the remarks for the previous example of a quadratic polynomial case, we could just deal with the case of a single third-degree term and make use of the results of the previous example and apply the principle of super-position. However, we will consider the function as shown in (5.122) as our starting point. Reference in made to Figure 5.5 (see Section 5.4).

By direct integration of (5.122), we have the reaction

$$R = af + \left(\frac{b}{2}\right)f^2 + \left(\frac{c}{3}\right)f^3 + \left(\frac{g}{4}\right)f^4 \tag{5.123}$$

For the region $0 < x < e$, the bending moments are

$$M_1 = R(X - x) \tag{5.124}$$

where

$$X = \frac{M_0}{R} \tag{5.125}$$

with

$$M_0 = K + eR \tag{5.126}$$

$$K = \left(\frac{a}{2}\right)f^2 + \left(\frac{b}{3}\right)f^3 + \left(\frac{c}{4}\right)f^4 + \left(\frac{g}{5}\right)f^5 \tag{5.127}$$

Therefore,

$$M_1 = K + (e - x)R \tag{5.128}$$

For the region $e < x < e + f$

$$M_2 = A - yB \tag{5.129}$$

where

$$A = K - k(t), \qquad B = R - r(t) \tag{5.130}$$

with

$$k(t) = \left(\frac{a}{2}\right)t^2 + \left(\frac{b}{3}\right)t^3 + \left(\frac{c}{4}\right)t^4 + \left(\frac{g}{5}\right)t^5 \tag{5.131}$$

$$r(t) = at + \left(\frac{b}{2}\right)t^2 + \left(\frac{c}{3}\right)t^3 + \left(\frac{g}{4}\right)t^4 \tag{5.132}$$

It is interesting to note that $k(f) = K$ and $r(f) = R$, where $k(f)$ and $r(f)$ are values of the functions $k(t)$ and $r(t)$ respectively, all evaluated at $t = f$. Thus

$$M_2 = K - k(t) + tr(t) - tR \tag{5.133}$$

These functions $k(t)$ and $r(t)$ are, in essence, characteristic functions derived from the loading function. As far as the region defined by $0 < x < e$ is concerned, M_1 is completely determined by K, R, e, and x. With a given load intensity function and location and extent of the loaded region, M_1 is a very simple linear function of x. Under the same conditions, the bending moments in the loaded region, namely M_2, are quite different and are fifth-degree polynomials of the location of the section under consideration. For the region $x > e + f$, $M_3 = 0$.

Now let us look at the deflection Y at the tip. Y can be written as the sum of three terms. Thus,

$$Y = Y_1 + Y_2 + Y_3 \tag{5.134}$$

with JY_i from M_i ($i = 1, 2, 3$) as

$$JY_1 = e(LM_0) - e^2 \frac{LR + M_0}{2} + e^3\left(\frac{R}{3}\right) \tag{5.135}$$

$$JY_2 = FG - N \tag{5.136}$$

where F is as shown in Figure 5.4 and

$$G = Kf - \left(\frac{R}{2}\right)f^2 + \left(\frac{a}{6} + \frac{bf}{24} + \frac{cf^2}{60} + \frac{gf^3}{120}\right)f^3 \tag{5.137}$$

$$N = \frac{Kf^2}{2} - \frac{Rf^3}{3} + f^4\left(\frac{a}{8} + \frac{bf}{30} + \frac{cf^2}{72} + \frac{gf^3}{140}\right) \tag{5.138}$$

and as was defined previously, $J = EI$. Note that $Y_3 = 0$, because $M_3 = 0$.

From Expressions (5.125) and (5.135), we see that JY_1 is the product of the reaction R, the distance e, and a quadratic polynomial in e. Similarly, from Expressions (5.136–5.138), it can be shown easily that JY_2 is the product of the extent f of the loaded area and a sixth-degree polynomial in f. The coefficients in these polynomials involve reaction R, span length L for JY_1, and the entities F and R for JY_2, in addition to the given coefficients a, b, c, and g in the load intensity function for both JY_1 and JY_2.

We can rewrite G as

$$G = \frac{f^3}{2}\left[\frac{a}{3} + \left(\frac{b}{4}\right)f + \left(\frac{c}{5}\right)f^2 + \left(\frac{g}{6}\right)f^3\right] \tag{5.139}$$

Similarly, we can rewrite N as

$$N = \frac{f^4}{6}\left[\frac{a}{4} + \frac{bf}{5} + \frac{cf^2}{6} + \frac{gf^3}{7}\right] \tag{5.140}$$

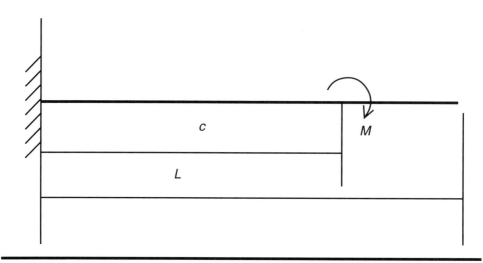

FIGURE 5.5. A concentrated couple on a cantilever

5.4. A CONCENTRATED COUPLE AT AN ARBITRARY POINT ON THE SPAN

The cantilever with a concentrated couple M_0 acting at an arbitrary point on the span is depicted in Figure 5.5. The end moment is equal to M_0. The bending moments for the two regions are as follows. For region $0 < x < c$

$$M = M_0 \tag{5.141}$$

For region $x > c$

$$M = 0 \tag{5.142}$$

The deflections Y are

$$Y = -(M_0)\,\frac{x^2}{2J} \quad \text{for } 0 < x < c \tag{5.143}$$

$$Y = -(M_0)c\left(\frac{x - \dfrac{c}{2}}{J}\right) \quad \text{for } x > c \tag{5.144}$$

provided we take downward deflection as positive and use the usual beam sign convention. In other words, the applied moment as shown in Figure 5.5 is negative in value, which, with the minus sign in Formulae (5.143) and (5.144), will render the deflections positive.

The deflection at the point of application of the concentrated couple is

$$Y = \frac{-(M_0)c^2}{2J} \tag{5.145}$$

The deflection at the tip of the cantilever is

$$Y = \frac{(-M_0)c\left(L - \dfrac{c}{2}\right)}{J} \tag{5.146}$$

Some interesting special cases are as follows.

Special Case 1. c = L

This is the case where the concentrated couple is at the tip of the cantilever. The bending moment is $M = M_0$ for the entire beam.

The general deflection formula is

$$Y = \frac{-(M_0)x^2}{2J} \quad \text{for } 0 < x < L \tag{5.147}$$

At the tip, the deflection is

$$Y = \frac{-(M_0)L^2}{2J} \tag{5.148}$$

Note that from (5.144), Y is proportional to M_0 and $c(x - c/2)$. It is interesting to probe this a little further, so let us consider $Y = -(M_0/J)\, c(x - c/2)$ as a function of c by setting

$$JY' = -(M_0)x + (M_0)c = 0 \tag{5.149}$$

We have

$$c = x \tag{5.150}$$

as the critical point. Formula (5.150) means that when the location of the point of application of the concentrated couple coincides with the section under consideration, Y has an extreme value. Now from (5.149) and the given condition about M_0, we obtain

$$JY'' = M_0 < 0 \tag{5.151}$$

Thus, we have a maximum value of Y at this critical point:

$$\max Y = \frac{-(M_0)x^2}{2J} \tag{5.152}$$

It is very interesting to note that Formula (5.152) looks exactly the same as Formula (5.147) and even Formula (5.143). Upon a moment's reflection while looking at (5.150), it is not surprising at all. Let us look at another special case.

Special Case 2. c = L/2

Here we have

$$M = M_0 \quad \text{for } 0 < x < \frac{L}{2} \tag{5.153}$$

$$M = 0 \quad \text{for } x > \frac{L}{2} \tag{5.154}$$

$$Y = \frac{-(M_0)x^2}{2J} \quad \text{for } 0 < x > \frac{L}{2} \tag{5.155}$$

$$Y = (-M_0)L \frac{x - \dfrac{L}{4}}{2J} \quad \text{for } x > \frac{L}{2} \tag{5.156}$$

$$Y = \frac{-(M_0)L^2}{8J} \quad \text{at } x = \frac{L}{2} \tag{5.157}$$

and

$$Y = -3(M_0)\left(\frac{L^2}{8J}\right) \tag{5.158}$$

at the tip of the cantilever.

It is interesting to observe that the deflections at the tip for the cases where $c = L$ and $c = L/2$ are related in the following manner. The ratio of the former to the latter is 4 vs. 3 only.

Let us compare the deflections at the mid-span, Y_1, and at the tip, Y_2, for the two special cases cited here. For the case where $c = L$, we have $Y_1/Y_2 = 1/4$. For the case where $c = L/2$, we have $Y_1/Y_2 = 1/3$.

5.5. EXPLORATIONS AND OBSERVATIONS

There are some interesting features about deflections of a cantilever under various loads when we compare results from these problems.

Let us start with Section 5.4, where the expressions for deflections are the simplest. The general expression for the first region defined by $0 < x < c$ is a single-term quadratic function of the location of the beam section under consideration, namely x. The general formula for the region defined by $x > c$, henceforth called the second region, is a two-term linear function of the location of the beam section. This linear function, expressed in Formula (5.144) of Section 5.4, represents a straight line which is a tangent to the second-degree curve depicting the deflection for the first region. The point of tangency is, of course, the point of application of the concentrated couple. Moreover, this linear function can be viewed as the product of three factors: M_0, $(x - c/2)$, and c/J. The effects of each one of these factors on the deflection are obvious to see. Even the effect of x is also very clear, and this fact can open up another viewpoint about the same formula.

This second viewpoint is to treat the deflection formula as the sum of two terms. One term is $(cM_0)x/J$, which is proportional to four factors: the location of the beam section, that of the point of application of the concentrated couple, and the concentrated couple itself, as well as $1/J$. The other term is $-(M_0)c^2/(2J)$, which has the obvious factors and is an entity independent of x. As a result, this constant term (relative to x) can be interpreted as the contribution to the deflection in the second region from a source in the first region. The last point is of fundamental importance when we are linking several problems under different loads in an effort to find common ground for these seemingly very different problems.

Two examples illustrating the application of the general load intensity function were presented in Section 5.3. One problem dealt with a general quadratic load intensity function and the other treated a general third-degree load intensity function. Deflection at the tip of the cantilever for each of the two examples is expressed as the sum of two terms, Y_1 and Y_2. This is done naturally with the added advantage of clarity. Here, Y_1 represents the contribution to the deflection from region $x < e$, the first region, whereas Y_2 is the contribution from the region defined by $e < x < e + f$, the second region.

For the first example, Y_2 is a sixth-degree polynomial in f, the extent of the loaded region, and is the product of f and a fifth-degree polynomial in f. The F_i's can be positive, negative, or zero, depending on the given coefficients in the load intensity function. Moreover, F_1 and F_2 have something in common in that they both have R as a factor.

For the second example, Y_2 is a seventh-degree polynomial in f and is the product of f and a sixth-degree polynomial in f. The highest and lowest order terms in Y_2 for $K \neq 0$ and $f \neq 0$ are nonzero, and the other terms may be positive, negative, or zero, depending on the coefficients in the load intensity function. Here, for Y_2, both examples have the common factor f.

We described Y_2 in detail in both problems. What about Y_1? It may be surprising to learn that Y_1 for both problems is exactly the same formally. It is a third-degree polynomial in e, the distance between the face of support and the beginning of the loaded region when we realize that R, X, and M_0 are related by $M_0 = RX$ in Section 5.3. Here, Y_1 is the product of e, R, and a quadratic polynomial in e. This is similar to Y_2 in that in both cases Y_i contains a factor of the independent variable. From Formula (5.72) in the discussion of the development of general load intensity function in Section 5.3, we see that the contribution to deflection from M_1 in the first region is more direct than from M_2 in the second region, at least formally for any load intensity function. "Formally" here means mathematically as well as physically. Of course, each problem has its own reaction, bending moments, etc., and the reaction in one problem is not equal to that in another one. However, a reaction is a reaction, no matter what load intensity function we are dealing with, as long as the universe of discourse is the same. The same remarks apply to Formulae (5.75) and (5.76) in the same section for any point in the second and third regions respectively.

6

EXAMPLES OF BEAM FORMULAE:
EXPLORATIONS AND COMMENTARY

6.1. INTRODUCTION

The purpose of this chapter is to exhibit collectively, in sufficient detail, more examples that illustrate the exploration of beam formulae for revealing the physical significance of solution results. The approach to the implementation of this project is as follows. First of all, a formula to be investigated will be chosen. This formula is an equality that expresses an entity as a function F of several other relevant entities V_i that will impact the value of the function. Then one of these entities V_i will be selected as the parameter and we will try to determine the effects of varying only this parameter on the value of F. After we finish the investigation of one such parameter, we can do another one. In each of these investigations, we are dealing, at any given time, with a real-valued function of one real variable only.

6.2. UNIFORM LOAD ON ONE SPAN OF
A TWO-SPAN CONTINUOUS BEAM: GENERAL CASE

Let us look at the general formula for the reaction R at the exterior support of a loaded span. From Section 3.1.4, we have

$$R = -\frac{\dfrac{L_1 L_2 A}{J_1} + \dfrac{3B}{J_2}}{D} \qquad (6.1)$$

For convenience and efficiency of subsequent operations, let us define the numerator in Formula (6.1) by the symbol C and recall the definition of D presented in Section 3.1.4. Reference is also made to Figure 3.3 in Section 3.1.4. Thus,

137

$$R = \frac{C}{D} \tag{6.2}$$

Let us examine the ingredients of C first. We see L_1, L_2, J_1, J_2, A, and B on the surface. Upon further inspection, we know from this list that only L_i and J_i are fundamental entities. A and B depend on other entities which are fundamental. Both A and B contain b, c, and L_1 in addition to the common factor wb.

Now we will look at D to find out what it is made of. D contains L_1, L_2, J_1, and J_2. Note that in D, the span length L_1 appears only once, whereas the span length L_2 appears in many places. The same is true in the case of C.

6.2.1. Parametric Study of J_1

To begin the parametric study, let us consider J_1 as the parameter. For convenience and uniformity in the use of symbols, we will call the parameter under study z. Therefore, R is now a function of the variable z. In order to gain more insight into the heart of the matter, let us proceed in two directions. First, we put R in the form

$$R = \frac{\left(\dfrac{e}{z}\right) + f}{\left(\dfrac{g}{z}\right) + h} \tag{6.3}$$

where e, f, g, and h are constants with respect to z. The expression shown in (6.3) is the result of direct "translation," in the literary sense, of Formula (6.1) with application of the definition of D. Now R in Formula (6.3) can be written as

$$R = \left(\frac{1}{h}\right)\left[f + \frac{G}{z + \left(\dfrac{g}{h}\right)}\right] \tag{6.4}$$

with

$$G = e - \frac{fg}{h} \tag{6.4A}$$

It can be seen clearly that the absolute value of the second term in the brackets in the expression for R decreases as z increases. It is also a simple matter to see that this second term itself, namely $G/(z + g/h)$, is negative, zero, or positive when G is negative, zero, or positive respectively. Therefore, R is a decreasing or increasing function of the independent variable z when G is positive or negative respectively. Note that both G and z play important roles here.

What about $G = 0$? Let us obtain the first derivative of R with respect to z as

$$R' = -Gh(hz + g)^{-2} \tag{6.5}$$

From (6.5), we see that G and R' have opposite signs; $G < 0$ or $G > 0$ corresponds to an increasing or decreasing R respectively. Also, $R' = 0$ if and only if $G = 0$, since the other factor in R' shown in (6.5) is not equal to zero. We see that $G = 0$ gives us the critical point. Furthermore, we have

$$R'' = (2Gh)^2(hz + g)^{-3} \tag{6.6}$$

Note that $h \neq 0$.

Thus, $R'' = 0$ if and only if $G = 0$ too. Therefore, $G = 0$ gives us a point of inflection. This is one direction.

Another direction is to proceed directly from the given form of representation shown in (6.1). We have the following.

Differentiating R with respect to z, we obtain

$$\frac{dR}{dz} = \frac{D\left(\dfrac{dC}{dz}\right) - C\left(\dfrac{dD}{dz}\right)}{D^2} \tag{6.7}$$

From (6.7), we see clearly that $dR/dz < 0$, $dR/dz = 0$, and $dR/dz > 0$ when the numerator is negative, zero, or positive respectively. This is because of the fact that the denominator in (6.7), namely D^2, is positive. For convenience, we will call the numerator in (6.7) N. Upon finishing the operations indicated in the definition of N, we arrive at the simple result

$$N = (L_2^2 A - 3B)(L_1 L_2^2)(J_1^2 J_2) \tag{6.8}$$

Focusing on (6.8), we see that N is negative, zero, or positive when the expression $(L_2^2 A - 3B)$, henceforth called K for brevity, is negative, zero, or positive respectively. This is due to the fact that L_1, L_2, J_1, and J_2 are all positive and $(L_1 L_2^2)/(J_1^2 J_2)$ is therefore positive also. Recall that A and B contain wb as a common factor in addition to the fact that they both contain b, c, and L_2 in the remaining portion. Thus, the characteristics of K, which indicate whether K is negative, zero, or positive, depend on b, c, and L_2 and how they interplay.

The conclusion is the following. R is a decreasing or increasing function of z when K is negative or positive respectively. R might have a maximum value or minimum value, or neither, when K is zero. As it turns out, R has neither a maximum nor a minimum value when K is zero, the critical point. Here, in the case at hand, $K = 0$ corresponds to an inflection point, as was indicated above when we dealt with one form of representation of R.

In both of these two approaches, we laid out the framework for our probing endeavor. We want to know what relationships there are among the entities b, c, and L_2 in order for K to be negative or positive in value or zero. Thus, we substitute A and B with their definitions and do the indicated mathematical operations. As a result of this, we obtain

$$K = (m - nL_2^2)wb \tag{6.9}$$

with

$$m = \frac{18bc^2 + 4b^2c + 12c^3 - 3b^3}{24} \tag{6.10}$$

$$n = \frac{\left(c + \dfrac{b}{2}\right)}{2} \tag{6.11}$$

From (6.9), it is clear that m, n, and L_2 completely determine whether K is negative, zero, or positive. That is,

$$K < 0 \text{ when } (m - nL_2^2) < 0 \tag{6.12}$$

$$K = 0 \text{ when } (m - nL_2^2) = 0 \tag{6.13}$$

$$K > 0 \text{ when } (m - nL_2^2) > 0 \tag{6.14}$$

For a given set of values of (b, c), we can get a corresponding set of (m, n) values and a corresponding value of L_2 in accordance with (6.13) or values of L_2 according to (6.12) or (6.14) depending on the given K value. Or, if we want, we can pick up any two of the three entities b, c, and L_2 as the given entities and determine the third one as a result. This is a good exercise for the reader to have some fun with the problem.

Case 1. The Critical Point K = 0

The immediate consequence is $AL_2^2 = 3B$ from the definition of K. Let us substitute this into the expression for R. We have the simple result

$$R = \frac{-A}{L_2} \tag{6.15}$$

In (6.15), since A is nonzero and negative and L_2 is positive, it follows that R is nonzero and positive. It is very interesting to note that R here, at least on the surface, depends on b, c, L_2, and w only. This is because of the fact that the entity A depends on these same entities only. This means that all the other entities including J_1, J_2, and L_1 do not have any influence on R at all. Here, the effects of b, c, w, and L_2 on R are obvious and can be read off from the expression for A. The reaction R is an increasing function of b and w as well as c, but is a decreasing function of the span length L_2. As mentioned in the statement immediately following (6.14), there are many possible values of (b, c, L_2) that satisfy (6.13). For each of these possible sets of values, we have an R that is given by (6.15). This last statement has far-reaching implications when $K = 0$, as will be seen regarding the relation among b, c, and L_2. Thus, if we put $L_2 = (m/n)^{1/2}$ in (6.15), then it can be seen that R depends on b, c, and w only, while L_2 disappears altogether from the expression for R. Furthermore, the other physical meaning derived from $K = 0$ is that R does not depend on J_i, along with some other entities mentioned above.

Case 2. K < 0

This is the case when (6.12) holds. There are many values of m that satisfy this. A sufficient condition for (6.12) to be true is $m < 0$. Another sufficient condition is $m = 0$. Both are interesting in that a rather large c relative to b is required just by observation. Of course, $m = 0$ and $m < 0$ are not necessary conditions for (6.12) to hold. The meaning of $K < 0$ has been mentioned above and will not be repeated here.

Case 3. K > 0

See Section 6.2.4 for details.

6.2.2. Parametric Study of J_2

If we observe the functional form of R as a real-valued function of a real variable J_2 and compare it with that, viewed as a real-valued function of the other real variable J_1 in the previous section, we will find that these two functions are similar. In fact, there is a certain degree of "symmetry" in a sense. We follow the same line of reasoning and the same approach to the exploration and come to the conclusion that the role played by K in Section 6.2.1 is now played by $-K$ in this section. All the results derived from K (for example, assertions based on K in Section 6.2.1) can be so interpreted. In other words, all the statements in Section 6.2.1 hold true here if K is replaced by $-K$. In particular, for the case where $K = 0$, all the statements in Section 6.2.1 remain unchanged in this section. It goes without saying that the K mentioned here is one and the same K as in Section 6.2.1.

6.2.3. Parametric Study of L_1

If K is so powerful or even magical, will we meet this character again? The answer is yes. In the study of L_1 here, the symbol $-K$ plays the role of K in Section 6.2.1.

In other words, as far as the effects on the functional behavior of R are concerned, J_1 and L_1 have something in common.

With obvious signs of the prominent importance of K appearing here and there, we are attempting to probe more and deeper into K and trying to uncover further detail, if any. For example, how much room do we have to expand our projects of exploration, and what special cases are there that should be addressed? At the same time, however, we are equally eager to see if the magical K will appear in more places, like in the study of C_1 at least. Thus, let us now proceed to the study of C_1 to find out.

6.2.4. Parametric Study of C_1 and the Marvel of the Kernel K

We just predicted that the marvelous K would show up again. Here it is already, with $-K$ here playing the role of K in Section 6.2.1. The conclusions to be reached here are obvious in view of what we did in Sections 6.2.2 and 6.2.3. In this connection, the reader may wonder what the reaction R is when $K = 0$ in Sections 6.2.2 and 6.2.3 and here. The answer is that in all these sections, as long as $K = 0$, we invariably have the result

$$R = \frac{-A}{L_2} \tag{6.15}$$

In terms of the fundamental entities, we can write R in (6.15) as

$$R = \left[L_2 - \left(c + \frac{b}{2} \right) \right] \frac{wb}{L_2} \tag{6.16}$$

The reader may notice the striking resemblance of the expression in (6.16) to a very familiar formula for a reaction for a simple beam problem. This is very interesting. If this is amazing, what about the vast number of possible choices of sets of (b, c, L_2) values that are available under the condition that $K = 0$ for our exploration? This is only for the case where $K = 0$ and we already have huge space for our projects to show their mighty power. We can consider the case where $K < 0$ and devise a b-by-c matrix with a controlling set of L_2 values and carry out interesting and useful projects. There is yet another way to explore and that is to pick one of the three entities (b, c, L_2) to be the parameter while holding the other two fixed. Similarly, we can treat the case where $K > 0$ with equal ease and efficiency.

Now let us see what values of the entities (A, B, L_2) or (b, c, L_2) there are to make the entity K very special and interesting. This will affect, of course, all cases where $K = 0$, $K < 0$, and $K > 0$.

Reference is made to Figure 3.3 in Section 3.1.4. If $a = 0$, then

$$b + c = L_2, \qquad m = (L_2)^2 \, \frac{2c - b}{4} + (3L_2 + c) \, \frac{b^2}{24},$$

$$n = \frac{L_2 + c}{4}$$

(6.17)

If $c = 0$, then

$$m = -\frac{b^3}{8}, \qquad n = \frac{b}{4}$$

(6.18)

If $a = 0$ and $c = 0$, then

$$b = L_2, \qquad m = \frac{-L_2^3}{8}, \qquad n = \frac{L_2}{4}$$

(6.19)

Next, let us look at the individual cases as follows.

Case 1. K < 0

Since A is negative and L_2 is positive, we have that AL_2^2 is negative, and the expression for K and the condition $B > 0$ will make K negative. Thus, $B > 0$ is a sufficient but not a necessary condition for $K < 0$. It is very interesting to note that $B > 0$ will never make K zero just by observing the definition of K.

Case 2. K = 0

The interesting thing about B is that here it is necessary that $B < 0$. A proof is as follows.

Proof

When $K = 0$, we must have either (a) $B < 0$ or (b) $B = 0$. This is because (c) $A < 0$, $L_2 > 0$, and the definition of K. However, $K = 0$ and $B = 0$ simultaneously means $A = 0$, which is contrary to the given condition (c), so (b) is not possible. Therefore, (a) holds.

Case 3. K > 0

Here again, B is an active player. $B < 0$ is a necessary but not a sufficient condition, just as in the case where $K = 0$.

Another interesting question is whether R can be zero or even negative. The conjecture is the negative answer. But can we prove it? It is definitely worthwhile to investigate. One of the best places to start is Formula (6.4). The interested reader is invited to pursue the task.

Is this the end of the magic power of K? It is left to the interested reader to find out.

6.3. TRIANGULAR LOAD ON ONE SPAN OF A TWO-SPAN CONTINUOUS BEAM: GENERAL CASE

The treatment here will be similar to that in the last section. We will take the reaction R at the exterior support of a loaded span for the subject of study. This reaction R was given in Section 3.1.8 as

$$R = -\frac{\dfrac{L_1 L_2 A}{J_1} + \dfrac{3B}{J_2}}{D} \tag{6.20}$$

Upon examining this formula and the formula for R in the case of a uniform load, we see that they are identical in form. The only difference is the actual expressions for the entities A and B between the two problems. However, even though there is this difference, the entities A and B in the two problems have something in common too. Here, both A and B contain b, c, w, and L_2 only, as in the previous problem. Moreover, as far as the case of studying the parameter L_2 is concerned, A has the same type of function here as in the previous section. This is also true of the situation with the entity B.

It is interesting to look at the exact content of m and n in the expression for K. Here

$$m = \frac{20bc^2 + 10c^3 + 5b^2c + 4b^3}{40} \tag{6.21}$$

$$n = \frac{2b + 3c}{12} \tag{6.22}$$

Note that the expression for m is very different from that for the uniform load problem. First of all, m here is never negative because every term in (6.21) is non-negative. Some terms may be zero, for example when $c = 0$, but never will any term become negative. This means that B is negative, due to the fact that B and m have opposite signs by definition. As a result of this, all the assertions in the last section concerning $B < 0$ will be applicable in the problem of triangular load here. Thus, the task of exploration is much simpler here for the theme in this section than in the last section where pioneering work in this aspect was done. We might even expect the same set of conclusions as in the last section in view of the many statements made at the beginning of this section which are identical to the corresponding ones in the last section. However, there are things that are unique to the triangular load problem. Some interesting special cases regarding m and n follow.

Special Case 1. c = 0

From Formula (6.22), we have the simplified

$$m = \frac{b^3}{10} \tag{6.23}$$

$$n = \frac{b}{6} \tag{6.24}$$

The kernel becomes

$$K = \left[\frac{-L_2^2 b}{6} + \frac{b^3}{10} \right] wb \tag{6.25}$$

Here, $K = 0$ when

$$3b^2 = 5L_2^2 \tag{6.26}$$

$K > 0$ when

$$3b^2 > 5L_2^2 \tag{6.27}$$

When we use K directly in association with the study of a particular parameter that calls for K in order to determine whether the function is increasing or decreasing or has a stationary value, it is important to pay attention to the following remarks.

Note that both Expressions (6.26) and (6.27) are inadmissible due to the very definition of L_2 and b. This means that neither $K = 0$ nor $K > 0$ will be admissible. Therefore, the only admissible condition for the kernel is $K < 0$. The requirement in terms of the relationship between b and L_2 under this condition is

$$3b^2 < 5L_2^2 \tag{6.28}$$

This is automatically satisfied by any b and L_2 and there are absolutely no restrictions whatsoever. When the parametric study calls for K as the direct indicator of the behavior of a function, $K < 0$ means that the reaction is a decreasing function of the parameter, for example J_1.

For those parametric studies which call for $(-K)$ instead of K to determine the behavior of a function in terms of whether it is an increasing or a decreasing function, we will just have to let $(-K)$ play the role of K. Thus, here we have that R is an increasing function of the parameter as the only admissible situation.

The principle for the special case where $c = 0$ is, then, that we only have $K < 0$ and no other values of K are possible. Moreover, no matter what the case may be, as long as $c = 0$ is the governing theme, K will never be zero. Once again, we see the fundamental importance of the kernel K.

Special Case 2. a = 0

We have, by definition, $L_2 = a + b + c$. Thus, $a = 0$ means $L_2 = b + c$. As a result, we can obtain m, n as in the general case or can proceed to replace L_2 by $b + c$ in the expression for K. If we take the former approach, we may want to ask the following questions. First of all, is there anything new here compared to the general case—anything at all? If so, what is it? Next, when will the kernel be positive, zero, or negative, if at all possible?

For the kernel K to be positive, we must have

$$m > nL_2^2 \qquad (6.29)$$

This means that L_2 must satisfy the following condition:

$$L_2 < \left(\frac{m}{n}\right)^{1/2} \qquad (6.30)$$

Therefore, it appears that positive K is admissible if L_2 is small enough.

Similarly, for negative K, we need

$$m < nL_2^2 \qquad (6.31)$$

This means that L_2 must satisfy

$$L_2 > \left(\frac{m}{n}\right)^{1/2} \qquad (6.32)$$

Therefore, it seems that negative K is admissible if L_2 is big enough.

Now consider the case where $K = 0$. The requirement is

$$m = nL_2^2 \qquad (6.33)$$

This means that

$$L_2 = \left(\frac{m}{n}\right)^{1/2} \qquad (6.34)$$

Thus, $K = 0$ is admissible if condition (6.34) is satisfied; that is, when L_2 is of the right size and that particular size is for real. In other words, $K = 0$ if we could find some b, c that will render (6.34) valid. However, is the actual situation really so?

Let us take the other approach mentioned above. The result comes out to be

$$K = -\left(\frac{wb^2}{40}\right)(8b^2 + 55bc + 20c^2) \tag{6.35}$$

It is clear from looking at Expression (6.35), where b and c are nonnegative, that

$$K < 0 \tag{6.36}$$

Special Case 3. a = 0 and c = 0

This is a further simplification from the case where $c = 0$ by letting $b = L_2$. Thus, $m = L_2^3/10$ and $n = L_2/6$. Therefore, all the statements under the case where $c = 0$ apply here.

6.3.1. Parametric Study of J_1, J_2, L_1, and C_1

With the preparation work done thus far, it is a simple matter to take care of the exploration of J_1, J_2, L_1, and C_1. The conclusions regarding the effects of any one of these parameters on the reaction are exactly the same as the corresponding ones in the case of a uniform load.

6.3.2. Parametric Study of L_2

This case differs from the ones listed above mainly due to the complicated form of involvement of L_2 in the expression for the reaction under investigation. For example, L_2 appears in A and B in a unique way which makes it very special, as we will see below. For convenience and diversity in presentation, we will call $-R/(wb)$ as y and we will work with y in the following. Thus,

$$y = \frac{-R}{wb} \tag{6.37}$$

is a function of z, which stands for the parameter L_2.

Starting with

$$\frac{A}{wb} = a_1 z + a_0 \tag{6.38}$$

$$\frac{B}{wb} = b_3 z^3 + b_2 z^2 + b_0 \tag{6.39}$$

$$\frac{R}{wb} = \frac{-C}{D} \tag{6.40}$$

where

$$C = e_3z^3 + e_2z^2 + e_1z + e_0 \tag{6.40A}$$

with

$$e_3 = \frac{3b_3}{J_2}, \qquad e_2 = \frac{3b_2}{J_2 + C_1a_2}, \qquad e_1 = C_1a_0, \qquad e_0 - \frac{3b_0}{J_2} \tag{6.40B}$$

and

$$D = z^2(d_1z + d_0) \tag{6.41}$$

with

$$d_1 = \frac{1}{J_2}, \qquad d_0 = C_1 \tag{6.42}$$

we obtain

$$y = \left(\frac{1}{d_1}\right)\left[e_3 + \frac{N}{M}\right] \tag{6.43}$$

where

$$M = (z^3 + C_1J_2z^2) \tag{6.44}$$

$$N = \left(\frac{e_2 - e_3d_0}{d_3}\right)z^2 + e_1z + e_0 \tag{6.45}$$

In order to examine the behavior of $R/(wb)$, we make use of the properties of the function y. Thus, let us look at the first derivative of y with respect to z

$$y' = \frac{Q}{M^2} \tag{6.46}$$

where

$$Q = MN' - M'N = z(g_3z^3 + g_2z^2 = g_1z + g_0) \tag{6.47}$$

with

$$g_3 = \frac{e_3d_0}{d_1} - e_2 = -\frac{2b + 3c}{4J_2} < 0 \tag{6.48}$$

$$g_2 = -2e_1 = -C_1\frac{2b + 3c}{3} < 0 \tag{6.49}$$

$$g_1 = -\left[C_1(2b + 3c)\frac{d_0}{d_1} + 3e_0\right] \tag{6.50}$$

$$g_0 = \frac{-2e_0 d_0}{d_1} > 0 \tag{6.51}$$

Note that g_3, g_2, and g_0 are nonzero, but it is possible for g_1 to be zero under the condition

$$18m = C_1^2 J_2^2 (2b + 3c) \tag{6.52}$$

which is very interesting in its own right as we note that the entity m is our old friend that showed up in the kernel K above.

It is interesting to observe some of the attributes of the g's by introducing a new notation $F = (2b + 3c)$. Thus, g_3 is proportional to F/J_2; the entity g_2 contains F and C_1 only; g_1 depends on F, C_1, J_2, and m; and finally g_0 is proportional to mC_1.

Now let us return to the main stream of investigating the behavior of the function y with the aid of Q defined above. We know that y is an increasing function of the independent variable z when

$$Q > 0 \tag{6.53}$$

and y is a decreasing function of z when

$$Q < 0 \tag{6.54}$$

Finally, y is stationary when

$$Q = 0 \tag{6.55}$$

In the last case, it is possible that y has extreme value(s). Therefore, let us look at $Q = 0$.

Since Q defined by Expression (6.47) is the product of z and a third-degree polynomial in z, with coefficients g's, we know that Q must have, in addition to the one trivial root $z = 0$, three other roots including at least one real root. Let us call these three roots z_1, z_2, and z_3. Thus, Expression (6.55) means that

$$g_3(z - z_1)(z - z_2)(z - z_3) = 0 \text{ or } z = 0 \tag{6.56}$$

We can proceed to include all the real positive roots and check for possible extreme values of y by the standard procedure in the usual manner and obtain the results.

For the case where $Q < 0$, we examine

$$g_3(z - z_1)(z - z_2)(z - z_3) < 0 \tag{6.57}$$

and note that for the nonzero roots z_1, z_2, and z_3 with the property

$$z_1 < z_2 < z_3 \tag{6.58}$$

we may want to use Table 6.1 to facilitate our work in deciding the range of values to be considered. Similarly, we can treat the case where $Q > 0$ by using Table 6.2.

TABLE 6.1. Type I

$z - z_1$	$z - z_2$	$z - z_3$
>0	<0	<0
>0	>0	>0
<0	<0	>0
<0	>0	<0

TABLE 6.2. Type II

$z - z_1$	$z - z_2$	$z - z_3$
>0	<0	>0
>0	>0	<0
<0	>0	>0
<0	<0	<0

Furthermore, after simplification, we have

$$e_3 = \frac{-1}{2J_2} \tag{6.59}$$

$$e_2 = \frac{2b + 3c}{4J_2} - \frac{C_1}{2} \tag{6.60}$$

$$e_1 = C_1 \frac{2b + 3c}{6} > 0 \tag{6.61}$$

$$e_0 = \frac{-1}{40J_2}[20bc^2 + 10c^3 + 5b^2c + 4b^3] \tag{6.62}$$

It is very interesting to note the following. The entity m that we encountered before is related to e_0. In fact,

$$m = -e_0 J_2 \tag{6.63}$$

Another interesting thing is that when $N = 0$, we have

$$y = \frac{e_3}{d_1} = -\frac{1}{2} \tag{6.64}$$

That is to say, here $R = wb/2$.

It may appear that this is a rather innocent result by looking at a special case of all the possible values that N can take. However, this result has profound implications. For one thing, it says that, regardless of how the span length L_2 under investigation varies, the reaction R is always a constant $wb/2$, which happens to be the maximum possible value that R can attain, as long as certain conditions are satisfied. Now the key questions are: What are these conditions and what do they mean in terms of physical significance?

To facilitate further probing, let us write N as

$$N = f_2 z^2 + e_1 z + e_0 \qquad (6.65)$$

where

$$f_2 = e_2 - \frac{e_3 d_0}{d_1} = \frac{2b + 3c}{4J_2} > 0 \qquad (6.66)$$

Now $N = 0$ implies, for a given set of values of (f_2, e_1, e_0), that z satisfies either one of the following two conditions:

$$z = \frac{-e_1 + S^{1/2}}{2f_2} \quad \text{henceforth called } z_1 \qquad (6.67)$$

$$z = \frac{-e_1 - S^{1/2}}{2f_2} \quad \text{henceforth called } z_2 \qquad (6.68)$$

where

$$S = [e_1^2 - 4f_2 e_0] \qquad (6.69)$$

Of course, we want to check that the z's are admissible. First of all, let us see whether we have two real roots for $N = 0$. To do this, we need to see

$$S > 0 \qquad (6.70)$$

This means that

$$e_1^2 - 4f_2 e_0 > 0 \qquad (6.71)$$

But $e_1^2 > 0$ and $f_2 e_0 < 0$ are known. Therefore, $S > 0$ and we do have two real roots for $N = 0$.

Next, do we have two positive roots for $N = 0$? The answer is no. Why not? This is because $-e_1$ and $-S^{1/2}$ are all negative and f_2 is positive in the definition of z_2. Therefore, we have only one admissible root for $N = 0$, namely z_1, given by Expression (6.67).

Another way to look at $N = 0$ is, for a given z, to find the values of other entities in the expression for N that make $N = 0$ valid.

Having dealt with the case where $N = 0$, it is only natural to raise the question about the situations where $N < 0$ and $N > 0$. Let us take the case where $N < 0$ first. This happens when

$$f_2(z - z_1)(z - z_2) < 0 \qquad (6.72)$$

Note that we can omit f_2 in (6.72) because $f_2 > 0$. Expression (6.72) can be satisfied when $(z - z_2) > 0$ and $(z - z_1) < 0$ simultaneously.

Finally, for the case where $N > 0$, one of the following has to hold: (a) both factors $(z - z_1)$ and $(z - z_2)$ are positive or (b) both factors $(z - z_1)$ and $(z - z_2)$ are negative.

From the definition of y, we see that the sign of N is of crucial importance since the second term in the formula for the function y has N as the numerator with a positive denominator. In other words, the sign of N alone determines the sign of the entire second term in the expression defining y. Note, moreover, that $N < 0$ means an increase in the absolute value of R compared with cases where $N = 0$ and $N > 0$ when we look at the definitions of y and R. Thus, we gain insight into the behavior of R with the aid of N in addition to the method using Q mentioned above. Of course, both methods are closely related and each has its own merits.

6.4. A CONCENTRATED COUPLE AT AN ARBITRARY POINT ON A TWO-SPAN CONTINUOUS BEAM: GENERAL CASE

A view of the expression for the reaction at the exterior support of a loaded span reveals that it is of a much simpler form than the ones in the last two sections. From Section 3.1.12, we have

$$R = \frac{AM_0}{D} \tag{6.73}$$

where

$$A = 3\left[\frac{C_1 L_2}{3} + \frac{L_2^2 - c^2}{2J_2}\right] \tag{6.74}$$

$$D = L_2^2(C_1 + C_2) \tag{6.75}$$

We see that the entity A depends on C_1, L_2, J_2, and c only.

Let us proceed to the parametric study.

6.4.1. Parametric Study of J_1

For convenience in subsequent development, we will call R/M_0 as y and denote the parameter under study as z. Then we will express y as a function of z in the following manner

$$y = \frac{\dfrac{a_1}{z} + a_0}{\dfrac{d_1}{z} + d_0} \tag{6.76}$$

where

$$a_1 = L_1 L_2, \qquad a_0 = \left(\frac{3}{2}\right)\frac{L_2^2 - c^2}{J_2} \tag{6.77}$$

$$d_1 = L_1 L_2^2, \qquad d_0 = L_2^2 C_2 \tag{6.78}$$

We can rewrite y as

$$y = \left(\frac{1}{d_0}\right)\left[a_0 + \frac{N}{z + \dfrac{d_1}{d_0}}\right] \tag{6.79}$$

where

$$N = a_1 - \frac{a_0 d_1}{d_0} \tag{6.80}$$

Then, we have the first derivative of y with respect to z:

$$y' = \frac{-N}{d_0}\left(z + \frac{d_1}{d_0}\right)^{-2} \tag{6.81}$$

It is clear from (6.81) that N is the controlling factor regarding the behavior of y. That is,

$$y' < 0 \quad \text{when} \quad N > 0 \tag{6.82}$$

$$y' = 0 \quad \text{when} \quad N = 0 \tag{6.83}$$

$$y' > 0 \quad \text{when} \quad N < 0 \tag{6.84}$$

Now N can be rewritten as

$$N = \frac{L_1}{2L_2}[3c^2 - L_2^2] \tag{6.85}$$

Therefore,

$$N > 0 \quad \text{when} \quad [3c^2 - L_2^2] > 0 \tag{6.86}$$

$$N = 0 \quad \text{when} \quad [3c^2 - L_2^2] = 0 \tag{6.87}$$

$$N < 0 \quad \text{when} \quad [3c^2 - L_2^2] < 0 \tag{6.88}$$

The immediate conclusions from the above are as follows:

1. There is a critical span length for span 2, denoted by $L_0 = (3)^{1/2}c$, that determines whether a given L_2 will cause N to be positive, negative, or zero, which

in turn will decide whether y is a decreasing or increasing function of z or is stationary respectively.

2. $N < 0$ denotes a specific domain of L_2 values, for a given c, in which all the L_2 values will make y an increasing function of z.
3. $N > 0$ denotes a specific domain of L_2 values, for a given c, in which all the L_2 values will make y a decreasing function of z.
4. $N = 0$ means that when $L_2 = L_0$, function y will have stationary value(s).

If we look at the problem from a different viewpoint, we may want to hold L_2 constant and let c vary by doing analogous things to have a direct contrast with the four conclusions above. We may coin the terminology of a critical location of applied moment and let it play the active role. This way, we can have four more conclusions corresponding to the ones just listed.

It is interesting to look at the case where $N = 0$. The immediate consequence is a simplified expression for y. Hence, we have

$$y = \left(\frac{3}{2}\right) \frac{L_2^2 - c^2}{L_2^3} \tag{6.89}$$

which can be rewritten as

$$y = \frac{3c^2}{L_2^3} = \frac{1}{L_2} \tag{6.90}$$

because of the very definition of N and the given condition $N = 0$.

Let us return to the general case of N regarding its sign (whether it is positive, negative, or zero) and examine the special case where $c = 0$. We have

$$N = -\frac{L_1 L_2}{2} \tag{6.91}$$

and we are forced to accept the fact that once $c = 0$ is the case, it is necessary that $N < 0$, meaning $y' > 0$.

Let us look at the case where $c = L_2$. We have $N = L_1 L_2 > 0$, meaning $y' < 0$. That is to say the following: y is a decreasing function of z.

6.4.2. Parametric Study of J_2

Once again, we will denote R/M_0 by y and the parameter by z for convenience. Then, from Formula (6.73), we have the same forms of expression for y as shown in Formulae (6.76) and (6.79). However, the coefficients are different. Here we have

$$a_1 = \left(\frac{3}{2}\right)(L_2^2 - c^2), \qquad a_0 = C_1 L_2 \tag{6.92}$$

$$d_1 = L_2^3, \qquad d_0 = L_2^2 C_1 \tag{6.93}$$

Moreover,

$$N = \frac{L_2^2 - 3c^2}{2} \tag{6.94}$$

Regarding the behavior of y, we can state the following. Since

$$y' = -\left(\frac{N}{d_0}\right)\left(z + \frac{d_1}{d_2}\right)^{-2} \tag{6.95}$$

we see right away that

$$y' > 0 \text{ when } N < 0 \tag{6.96}$$

$$y' = 0 \text{ when } N = 0 \tag{6.97}$$

$$y' < 0 \text{ when } N > 0 \tag{6.98}$$

All these may look like mere repetition of Section 6.4.1. However, the N here is exactly the negative of the N there except for a positive factor. In other words, N here plays the role of $(-N)$ in Section 6.4.1.

Now let us look at the case where $N = 0$. Here, surprisingly enough, we get

$$y = \frac{1}{L_2} \tag{6.99}$$

just as we did in Section 6.4.1.

What about the special case where $c = 0$ when we are focusing on the general case of N to begin with, that is, regardless of whether N is positive, negative, or zero in the first place? As we might expect, the answer is simply $N > 0$, with

$$N = \frac{L_2^2}{2} \tag{6.99A}$$

and

$$y = \frac{1}{L_2} + \frac{1}{2C_1\left(z + \dfrac{L_2}{C_1}\right)} \tag{6.99B}$$

For the sake of satisfying our curiosity, let us look at the case where $c = L_2$. We get

$$N = -L_2^2 < 0 \tag{6.99C}$$

and

$$y = \frac{1}{L_2} - \frac{1}{C_1 z + L_2} \tag{6.99D}$$

6.4.3. Parametric Study of L_1

The expression for A can be rewritten as a function of the parameter under study, z, in the following manner

$$A = a_1 z + a_0 \tag{6.100}$$

where

$$a_1 = \frac{L_2}{J_1}, \qquad a_0 = 3 \frac{L_2^2 - c^2}{2J_2} \tag{6.101}$$

The expression for D can be rewritten as a function of z in the form

$$D = d_1 z + d_0 \tag{6.102}$$

where

$$d_1 = \frac{L_2^2}{J_1}, \qquad d_0 = L_2^2 C_2 \tag{6.103}$$

The function $y = R/M_0$ now takes the form

$$y = \left(\frac{1}{d_1}\right)\left[a_1 + \frac{N}{z + \dfrac{d_0}{d_1}}\right] \tag{6.104}$$

where

$$N = a_0 - \frac{a_1 d_0}{d_1} = \frac{L_2^2 - 3c^2}{2J_2} \tag{6.105}$$

Now

$$y' = \left(\frac{-N}{d_1}\right)\left[z + \frac{d_0}{d_1}\right]^{-2} \tag{6.106}$$

We see that

$$y' < 0 \text{ when } N > 0 \tag{6.107}$$

$$y' = 0 \text{ when } N = 0 \tag{6.108}$$

$$y' > 0 \text{ when } N < 0 \tag{6.109}$$

since all the factors except N in y' are positive, but capped with a minus sign.

For $N = 0$, we have

$$L_2^2 = 3c^2 \qquad (6.110)$$

and a simplified expression for y is

$$y = \frac{1}{L_2} \qquad (6.111)$$

All the remarks about N and critical span length, etc. made in Section 6.4.1 apply here except that N here plays the role of $(-N)$ there.

Two extreme cases regarding R are of special interest.

Case 1. c = 0

We obtain

$$N = \frac{L_2^2}{2J_2} > 0 \qquad (6.112)$$

signifying an increasing function for R.

Case 2. c = L_2

We have

$$N = \frac{-L_2^2}{J_2} < 0 \qquad (6.112A)$$

indicating a decreasing function for R.

6.4.4. Parametric Study of L_2

Following the standard procedure for setting up a parametric study, we will write A as a function of the parameter z in the form

$$A = a_2 z^2 + a_1 z + a_0 \qquad (6.113)$$

where

$$a_2 = \frac{3}{2J_2}, \qquad a_1 = C_1, \qquad a_0 = \frac{-3c^2}{2J_2} \qquad (6.114)$$

and we will write D as

$$D = z^2(d_1 z + d_0) \qquad (6.115)$$

where

$$d_1 = \frac{1}{J_2}, \qquad d_0 = C_1 \tag{6.116}$$

The function to be examined is denoted, as usual, as y and

$$y = \frac{A}{D} \tag{6.117}$$

Its first derivative is

$$y' = \frac{DA' - AD'}{D^2} = \frac{N}{z^3(d_1z + d_0)^2} \tag{6.118}$$

where

$$N = b_3z^3 + b_2z^2 + b_1z - m_0 \tag{6.119}$$

will be called the critical function for y for convenience in the following development, with

$$b_3 = -a_2d_1 = \frac{-3}{2J_2^2} < 0 \tag{6.120}$$

$$b_2 = -2a_1d_1 = \frac{-2C_1}{J_2} < 0 \tag{6.121}$$

$$b_1 = 3a_0d_1 - a_1d_0 = \frac{9c^2}{2J_2^2} - C_1^2 \tag{6.122}$$

$$m_0 = 2a_0d_0 = \frac{-3c^2C_1}{J_2} \tag{6.123}$$

As usual, the key to the problem of understanding the function y is to examine N, the critical function for y. Since N is a third-degree polynomial in general, we can follow the same procedure as we did in Section 6.4.1.

In order to pave the way for dealing with two very interesting and important cases, we will note the following features of the coefficients b_1 and m_0 in N. Note that $c = 0$ when $m_0 = 0$. For other values of c, we have $m_0 < 0$. As for the coefficient b_1, it may be positive, negative, or zero depending on

$$\frac{9c^2}{2J_2^2} - C_1^2 > 0, < 0, \text{ or } = 0 \tag{6.124}$$

respectively. The coefficient b_1 has the special property

$$b_1 = 0 \quad \text{when} \quad \frac{9c^2}{2J_2^2} - C_1^2 = 0 \tag{6.124A}$$

Both b_1 and m_0 are very important, and it is worthwhile to look into individual cases.

Case 1. $m_0 = 0$

When $m_0 = 0$, we have

$$N = (b_3 z^2 + b_2 z + b_1) z \qquad (6.125)$$

That is to say, N is a product of z and a quadratic expression N_1 in z, while y' is simplified to

$$y' = \frac{N_1}{[z(d_1 + d_0)]^2} \qquad (6.126)$$

where

$$N_1 = b_3 z^2 + b_2 z + b_1 \qquad (6.127)$$

Thus, y' is positive, negative, or zero when N is positive, negative, or zero respectively. In other words, the function y will be an increasing function, a decreasing function, or a function that has a stationary value depending on whether $N_1 > 0$, < 0, or $= 0$ respectively.

Let us look at N_1 closely for a moment. It is clear that if $b_1 < 0$, then because all the b's are negative and z is positive, we have $N_1 < 0$.

If $b_1 > 0$, then it is possible that $N_1 = 0$, < 0, or > 0. The case $N_1 < 0$ holds if

$$b_1 < -(b_3 z^2 + b_2 z) \qquad (6.128)$$

and the case $N_1 > 0$ holds if

$$b_1 > -(b_3 z^2 + b_2 z) \qquad (6.129)$$

Let us study the roots of $N = 0$. The nature of the roots depends, to a large extent, on the sign of b_1, the constant term in N_1.

If $b_1 > 0$, then N_1 has only one positive real root and it is

$$z = \frac{-b_2 - S}{2b_3} \quad \text{with} \quad S = (b_2^2 - 4b_1 b_3)^{\frac{1}{2}} \qquad (6.130)$$

If $b_1 < 0$, then we must first check to see whether $N_1 = 0$ has any real root(s) at all, whether positive or negative. We already know from a naïve look above. However, we could also do it differently. Thus, for real roots, we need to have

$$\frac{27 c^2}{J_2^2} \geq 2 C_1^2 \qquad (6.131)$$

However, even with this provision, we can only have negative roots and no positive root(s) at all.

Regardless of whether $b_1 < 0$ or $b_1 > 0$, we can use the standard procedure in the previous section regarding a critical quadratic equation and deal with the different domains of z describing an increasing or a decreasing function of z or a stationary value of the function.

If $b_1 = 0$, then we have another special case to deal with.

Case 2. $b_1 = 0$

This means that there is a specific nice relation among c, C_1, and J_2, namely:

$$\frac{(3c)^2}{J_2^2} - 2C_1^2 = 0 \tag{6.132}$$

Also, N becomes

$$N = b_3 z^3 + b_2 z^2 - m_0 \tag{6.133}$$

where m_0 becomes

$$m_0 = \frac{-2C_1^3 J_2}{3} \tag{6.134}$$

or, in terms of c and C_1, we have

$$m_0 = -2\left(\frac{3}{2}\right)^{1/2} cC_1^2 \tag{6.135}$$

or, in terms of c and J_2, we have

$$m_0 = -3\left(\frac{3}{2}\right)^{1/2} \frac{c^3}{J_2^2} \tag{6.136}$$

Case 3. $b_1 = 0$ and $m_0 = 0$

Is this possible? From the definitions of b_1 and m_0 we know that the answer is negative. This is due to the fact that $m_0 = 0$ requires $c = 0$, thereby giving b_1 a nonzero value.

6.5. A CONCENTRATED FORCE AT AN ARBITRARY POINT ON AN EXTERIOR SPAN OF A THREE-SPAN CONTINUOUS BEAM: GENERAL CASE

The reaction R_1 was given previously as

$$R_1 = (As + BK)\frac{P}{D} \tag{6.137}$$

where

$$A = L_1 L_3^2 C_2 \frac{C_2 + C_3}{18} \tag{6.138}$$

$$B = \frac{L_1 L_2 L_3}{6 J_2} \tag{6.139}$$

In order to raise the reader's interest and change pace, the material in this section will be presented in a different order.

6.5.1. Parametric Study of L_1

As usual, we will denote the parameter under study as z. Thus, we have

$$As = az \tag{6.140}$$

where

$$a = \left[C_2 L_3^2 (C_2 + C_3) \frac{s}{18} \right] \tag{6.141}$$

and

$$BK = bz \tag{6.142}$$

where

$$b = \left[\frac{L_2 L_3}{6 J_2} \right] K \tag{6.143}$$

with K independent of L_1.

Note that the coefficient a is an entity that consists of C_2, C_3, L_3, and s only, and the coefficient b is a product of K and a function of J_2, L_2, L_3, s, and a numerical factor.

The symbol D can have the following look

$$D = \frac{z^2 L_3^2}{36} [d_0 + d_1 z] \tag{6.144}$$

where

$$d_0 = C_2(3C_2 + 4C_3) \tag{6.145}$$

is a function of C_2 and C_3 only and

$$d_1 = \left(\frac{4}{J_1} \right) (C_2 + C_3) \tag{6.146}$$

is a function of C_2, C_3, and J_1 only.

With this introduction, we can rewrite R_1 as

$$R_1 = \frac{36(a + b)}{L_3^2 z(d_0 + d_1 z)} \tag{6.147}$$

We see that R is essentially a function, called y, of the variable z multiplied by a constant. Thus, let us examine a new function defined as

$$y = \frac{1}{z(d_0 + d_1 z)} \tag{6.148}$$

The first derivative of y is negative for all values of z and is never zero. This can be seen from the following:

$$y' = -[z(d_0 + d_1 z)]^{-2}[d_1 z + d_0] < 0 \tag{6.149}$$

Furthermore, for

$$y' = 0 \tag{6.150}$$

we must have

$$z = \frac{-d_0}{2d_1} < 0 \tag{6.151}$$

However, z is always positive; therefore, it is not possible that (6.150) and (6.151) hold.

Another way to look at the implication of y' is to note that the denominator of y' in all cases is positive, so that the sign of y' is determined completely by that of the numerator, which is $-(2d_1 z + d_0)$ and is negative. Therefore, y is a decreasing function of z. As a result of this and the definition of R_1, we see that R_1 is also a decreasing function of z.

6.5.2. Parametric Study of C_2

Here we have for the elements needed in R_1

$$A = a_2 z^2 + a_1 z \tag{6.152}$$

where

$$a_2 = \frac{L_1 L_3^2}{18} \tag{6.153}$$

$$a_1 = \frac{L_1 L_3^2 C_3}{18} \tag{6.154}$$

$$BK = b_1 z + b_2 z^2 \tag{6.155}$$

with

$$K = K_0 + zK_1 \tag{6.156}$$

$$b_1 = \frac{L_1 L_3 K_0}{6} \tag{6.157}$$

$$b_2 = \frac{L_1 L_3 K_1}{6} \tag{6.158}$$

Also,

$$D = \left[\frac{L_1 L_3}{6} \right]^2 [3z^2 + d_1 z + d_0] \tag{6.159}$$

with

$$d_0 = 4C_1 C_3, \qquad d_1 = 4(C_1 + C_3) \tag{6.159A}$$

Therefore, with R_1 given by (6.137) in mind, we have

$$\frac{As + BK}{D} = \frac{12}{(L_1^2 L_3^2)} [sa_2 + b_2 + y] \tag{6.160}$$

where

$$y = \frac{f}{g} \tag{6.161}$$

with

$$f = (mz + n) \tag{6.162}$$

$$g = (z^2 + ze_1 + e_0) \tag{6.163}$$

$$e_0 = \frac{d_0}{3}, \qquad e_1 = \frac{d_1}{3} \tag{6.164}$$

$$m = sa_1 + b_1 - \frac{(sa_2 + b_2)d_1}{3} \tag{6.165}$$

and

$$n = - \frac{(sa_2 + b_2)d_0}{3} \tag{6.166}$$

Considering y as a function of z, we have the first derivative of y as

$$y' = \frac{N}{g^2} \tag{6.167}$$

where

$$N = -mz^2 - 2nz + (me_0 - ne_1) = n_2z^2 + n_1z + n_0 \qquad (6.168)$$

We notice that the behavior of y and thus that of R_1 hinges on the nature of N, which in turn depends on the coefficients that appear in N. Therefore, it is both interesting and important to review m, n, e_0, and e_1 in this context.

Let us tackle m first. From the way in which m is defined, we see that m can take positive or negative values or can be zero. We will examine m in these three categories next.

Category A. m = 0

The immediate consequence is

$$y = \frac{n}{g} \qquad (6.169)$$

This is a much simpler result than in the cases where m is nonzero.

Note that the coefficients in the expression for g are all positive due to the nature and definition of e_0 and e_1. Also, recall that $n < 0$, so we see right away that y is negative. This means that R_1 may be zero or negative. The condition under which this happens is

$$b_2 + sa_2 + y = 0 \text{ or } < 0 \qquad (6.158A)$$

respectively.

From (6.169) and (6.158A) we have

$$b_2 + sa_2 + \frac{n}{g} = 0 \text{ or } < 0 \qquad (6.158B)$$

respectively.

The reaction R_1 may be zero or negative, but the first derivative of y is positive due to $n < 0$. In exact form, we have

$$y' = \frac{-n(2z + e_1)}{g^2} \qquad (6.170)$$

Let us ask the question: What does it mean to have $m = 0$ all by itself? From the definition of m, we have $m = 0$ when $e_i = Q$, where Q is defined as

$$Q = \frac{s(L_3C_3) + 3K_0}{sL_3 + 3K_1} \qquad (6.171)$$

This is a specific relation among e_1, C_3, L_3, s, K_0, and K_1. Recall that K is an expression that involves J_3, L_3, C_1, C_3, and s. Thus, it is clear that there are many different sets of values taken from the (C_i, J_3, L_3, s) domain where $i = 1, 3$ that will satisfy the condition specified in (6.171).

Formula (6.170) says that the bigger the value of $-n$ or $(2z + e_1)$ or their product becomes, the bigger the value of y' will get. This means that an increase in C_1, C_3, L_1, or L_3 translates into an increase in y'. In other words and in more precise terms, we can say that the rate of increase in y is proportional to the third power of L_3 and the square of L_1 and inversely proportional to the product of J_1 and J_3.

Now looking at Expression (6.169), we see that it is proportional to n, which is proportional to the product of L_1, $L_3^2 C_1$, and C_3. Thus, y is proportional to $L_1^2 L_3^3 / (J_1 J_3)$.

Category B. m < 0

This happens when $Q < e_1$. From the expression for y, we see immediately that the numerator is negative whereas the denominator remains positive. Thus, y is negative. Moreover, the bigger the absolute value of m is, the bigger that of y becomes. However, the situation with y' is not so obvious. We need to do more work. In other words, N could be positive, negative, or zero. For a given n, where n is negative, the key entity to be concerned with is the constant term in the expression for N, namely n_0. Note that n_0 could be positive, negative, or zero. Let us explore these cases one by one as follows.

Case 1. $n_0 = 0$. Here we have the simplified N as

$$N = z^2 n_2 + z n_1 \qquad (6.172)$$

which is positive for all $z > 0$ due to the fact that both coefficients in (6.172) are positive. Since the sign of N determines the sign of y', we see that here $y' > 0$ also. Thus, y is an increasing function of z.

Case 2. $n_0 > 0$. Next, here we have all positive coefficients in (6.172) and consequently $N > 0$ for positive z. The conclusion in this case is $y' > 0$. Again, y is an increasing function of z.

Case 3. $n_0 < 0$. Finally, here all the possibilities for N are open. First, $N = 0$ happens at

$$z_0 = \frac{-n_1 + (n_1^2 - 4n_0 n_2)^{1/2}}{2n_2} \qquad (6.173)$$

Note that even though $N = 0$ has two real roots due to the fact that $n_0 < 0$, one positive and one negative, only the positive root which is physically meaningful is presented here. It can be shown easily that the second derivative of y is positive and we have a minimum value of y at z_0.

Furthermore, the situation $N < 0$ occurs when $0 < z < z_0$. Here, we have the not so common situation where y is a decreasing function still in the category of $m < 0$.

Finally, of course, $N > 0$ exists when $z > z_0$.

Category C. m > 0

This category, which occurs when $e_1 < Q$ holds, is very similar to category B. Again, the basic and first thing to do is to look at n_0, the constant term in the expression for N. There are three cases to consider.

Case 1. $n_0 = 0$. Again, N is simplified to a two-term second-degree polynomial, but with $n_2 < 0$ and $n_1 > 0$. Hence, here it is possible to have $N = 0$. Thus, for $N = 0$

$$z_0 = \frac{-n_1}{n_2} \tag{6.174}$$

is the nontrivial root that has physical meaning. The second derivative of y is

$$y'' = -2Ng^{-3}g' + g^{-2}N' = g^{-2}(2n_2z + n_1) \tag{6.175}$$

by using the given condition $N = 0$.

Now y' could be positive, negative, or zero, unlike the situation in category B above. First, $y = 0$ when

$$(2n_2z + n_1) = 0 \tag{6.176}$$

that is, when

$$z = \frac{-n_1}{2n_2} \tag{6.177}$$

Next, $y'' > 0$ when

$$(2n_2z + n_1) > 0 \tag{6.178}$$

Finally, $y'' < 0$ when

$$(2n_2z + n_1) < 0 \tag{6.179}$$

Note that at z_0 we have $y'' < 0$, so y has a maximum at z_0 expressed in (6.174).

Case 2. $n_0 > 0$. The roots of $N = 0$ can be found in a similar form as in case 3 in category B. Thus,

$$z_0 = \frac{-n_1 - (n_1^2 - 4n_0n_2)^{1/2}}{2n_2} \qquad (6.180)$$

is the only admissible root of $N = 0$ in view of the fact that n_0 and n_2 have opposite signs in addition to the relative magnitude of the two terms in the bracket. At this value z_0, we have $y'' < 0$. Thus, y has a maximum value (algebraically, of course). Moreover, due to the given condition $n_2 < 0$, we have the following results: $N < 0$ when $z > z_0$. This means that y is a decreasing function of z when z is greater than z_0. $N > 0$ when $0 < z < z_0$. This means that y is an increasing function when z lies between 0 and z_0. For a nonnegative z, results are the same as $z < z_0$.

Case 3. $n_0 < 0$. $N = 0$ has two real roots when

$$S = n_1^2 - 4n_0n_2 > 0 \qquad (6.181)$$

It turns out that we have two positive roots, z_1 and z_2, here with $0 < z_1 < z_2$. The roots are

$$z_1 = (-n_1 + S^{1/2})(2n_2) \qquad (6.182)$$

$$z_2 = (-n_1 - S^{1/2})(2n_2) \qquad (6.183)$$

Moreover, y has a minimum at z_1 and a maximum at z_2. y is a decreasing function of z when $N < 0$, requiring $z > z_2$ or $z < z_1$. y is an increasing function when $N > 0$, which calls for $z_1 < z < z_2$.

Why we did so elaborately discuss all the different cases presented above? The reason is twofold: these cases exist under specific conditions relating to several parameters, and their combined action has very important physical significance.

6.5.3. Parametric Study of J_1

We notice that the expression for R_1 contains the entities A, B, and K, which are all independent of J_1; however, the entity D contains J_1, which will be called z below. Thus,

$$D = a + \frac{b}{z} \qquad (6.184)$$

where

$$a = \left[\frac{L_1L_3}{6}\right]^2 [C_2(3C_2 + 4C_3)] \qquad (6.185)$$

$$b = 4\left[\frac{L_1L_3}{6}\right]^2 L_1(C_2 + C_3) \qquad (6.186)$$

The reaction R_1 is

$$R_1 = \frac{C}{a + \dfrac{b}{z}} \tag{6.187}$$

where

$$C = As + BK \tag{6.188}$$

is a constant as far as the variable z is concerned.

For convenience in future work, let us write

$$R_1 = Cy \tag{6.189}$$

and consider y. We have the first derivative of y with respect to z as

$$y' = b(az + b)^{-2} > 0 \tag{6.190}$$

for all z because both factors of y' are positive. From (6.190), we know that y is an increasing foundation of z. Moreover, from (6.189), R is proportional to both c and y. We note that y' is not only nonzero for finite z, but its value drops very rapidly as z grows.

The second derivative of y is

$$y'' = -2ab\left[z\left(a + \frac{b}{z}\right)^{-3}\right] < 0 \tag{6.191}$$

because a and b are both positive. Note that the entities a, b, and C all have the common factors L_1, L_3. From (6.187), we see that this common factor is canceled out and the resulting expression for R_1 is thus simplified.

6.5.4. Parametric Study of J_3

As usual, let us denote the parameter as z. Here the entities A, B, D, and K are as follows. First, we have

$$A = a + \frac{b}{z} \tag{6.192}$$

where

$$a = \frac{L_1 L_3^2 C_2^2}{18} \tag{6.193}$$

$$b = \frac{L_1 L_3^2 C_2}{18} \tag{6.194}$$

Next we see

$$B = \frac{L_1 L_3 C_2}{6} \qquad (6.195)$$

Note that a, b, and B all have the common factors L_1, L_3, and $C_2/18$. Then, we notice that

$$D = c + \frac{d}{z} \qquad (6.196)$$

is similar in form to A, with

$$c = \left[\frac{L_1 L_3}{6}\right]^2 [C_2(3C_2 + 4C_1)] \qquad (6.197)$$

$$d = \left[\frac{L_1 L_3}{6}\right]^2 [(4L_3)(C_1 + C_2)] \qquad (6.198)$$

Finally

$$K = e + \frac{f}{z} \qquad (6.199)$$

is another entity that bears some resemblance to A, with

$$e = -\frac{sC_2 L_3}{3}, \qquad f = s^2 \frac{s - 3L_3}{6} \qquad (6.200)$$

Thus,

$$R_1 = \left(\frac{h}{z}\right)\left(c + \frac{d}{z}\right) \qquad (6.201)$$

where

$$h = bs + Bf \qquad (6.201A)$$

We can rewrite R_1 as

$$R_1 = \frac{h}{c\left(z + \frac{d}{c}\right)} \qquad (6.202)$$

Note that h needs not be positive all the time because of the presence of f, which may have the right value to render h zero or even negative on the one hand and may take other values that will result in a positive h on the other hand.

Now from (6.202), we have

$$R_i' = \frac{-\dfrac{h}{c}}{\left[z + \dfrac{d}{c}\right]^2} \tag{6.203}$$

The sign of R' depends on that of h because c is positive by definition. Therefore, let us work on h as follows.

We can rewrite h as

$$h = \frac{L_1 L_3 C_2}{36}\left(2L_3^2 - 3L_3 s + s^2\right) \tag{6.204}$$

First, we consider $h = 0$. This happens when

$$(2L_3 - s)(L_3 - s) = 0 \tag{6.205}$$

This means that either $2L_3 - s = 0$ or $L_3 - s = 0$. The former is physically impossible, so we are left with $L_3 - s = 0$. That is, $s = L_3$. This gives us $R_1 = 0$ from (6.202).

Next, we try $h < 0$. This occurs when

$$(2L_3 - s)(L_3 - s) < 0 \tag{6.206}$$

but this implies that these two factor must have opposite signs. Having $(2L_3 - s) > 0$ and $(L_3 - s) < 0$ simultaneously seems to be a good candidate. There is no problem with the former, but it goes against the grain to accept the latter when we reflect a little. Therefore, this choice does not work. Switching the signs around, of course, is not any better. Therefore, $h < 0$ is not possible.

We see immediately that $h > 0$ makes good sense. This requires

$$(2L_3 - s) > 0 \text{ and } (L_3 - s) > 0 \text{ simultaneously} \tag{6.207}$$

We already excluded $(2L_3 - s) < 0$ and $(L_3 - s) < 0$ simultaneously as candidates for consideration on the grounds of physical meaning. Let us return to (6.207) now. It is always the situation and is, therefore, satisfied automatically. We return to (6.203) and know that R_1' is negative when $h > 0$, so R_i is a decreasing function of z and is positive except when $h = 0$, which corresponds to $s = L_3$ and a zero-valued R_1.

6.5.5. Parametric Study of L_3

The ingredients for R_1, upon designating the parameter to be studied as z, are expressed as follows

$$A = a_1 z^2 + a_2 z^3 \tag{6.208}$$

with

$$a_1 = \frac{L_1 C_2^2}{18}, \qquad a_2 = \frac{L_1 C_2}{18 J_2} \qquad (6.209)$$

The entity B is given by

$$B = b_1 z \qquad (6.210)$$

with

$$b_1 = \frac{L_1 C_2}{6} \qquad (6.211)$$

The entity K is

$$K = k_0 + k_1 z \qquad (6.212)$$

with

$$k_0 = \frac{s^3}{6 J_3}, \qquad k_1 = \frac{-s}{6}\left[\frac{3s}{J_3} + 2C_2\right] \qquad (6.213)$$

Finally, the entity D is

$$D = \left(\frac{L_1}{6}\right)^2 z^2 (d_0 + d_1 s) \qquad (6.214)$$

where

$$d_0 = 3C_2^2 + 4C_1 C_2 \qquad (6.215)$$

$$d_1 = \frac{4}{J_3}(C_1 + C_2) \qquad (6.216)$$

As a result of this introduction, we obtain

$$R_1 = (6L_1)^2 \frac{f}{d_1} \qquad (6.217)$$

where

$$f = (az^2 + bz + c)\left(\frac{d_0 z}{d_1} + z^2\right) \qquad (6.218)$$

with

$$a = a_1 s > 0, \qquad b = a_1 s + b_1 k_1, \qquad c = b_1 k_0 > 0 \qquad (6.219)$$

The function f can be rewritten as

$$f = a + \frac{ez + c}{z^2 + mz} \tag{6.220}$$

with

$$m = \frac{d_0}{d_1} > 0, \quad e = b - ma \tag{6.221}$$

Let

$$f = a + y \tag{6.222}$$

Then, as can be seen from (6.222), f is the sum of a fractional function of z, designated y, and a constant.

Taking the first derivative of f, we have

$$f' = y' = \frac{N}{(z^2 + mz)^2} \tag{6.223}$$

where

$$N = -ez^2 - 2cz - mc \tag{6.224}$$

Note that the denominator in (6.223) is positive because each term is positive to begin with. Hence, the sign of y' is determined by that of N. Looking at the coefficients in N, we see that we have to take care of the sign of e first, since $c > 0$ and $m > 0$ from the way they were defined already. Let us do the work case by case as follows.

Case 1. $e > 0$

This happens when $b > am$ theoretically at best. But is this possible? The answer is negative. The reason is demonstrated below.

Proof

Suppose $e > 0$. Then, for given $a > 0$, $m > 0$, we have $ma > 0$. Since by definition $e = b - ma$, it is necessary to have $b > 0$ in order to achieve $b > ma$, meaning $e > 0$. This means

$$a_1 s > b_1 k_1 \tag{6.225}$$

Thus, we have

$$0 > \frac{3s}{J_3} \tag{6.226}$$

which is not true, because both s and J_3 are positive. Therefore, the supposition $e > 0$ is not valid.

Case 2. $e = 0$

Next we consider the possibility $e = 0$. This means

$$b = ma \tag{6.227}$$

but

$$b = \frac{-L_1 s^2 C_2}{12 J_3} < 0 \tag{6.228}$$

and

$$ma > 0 \tag{6.229}$$

Putting (6.227–6.229) together, we immediately see a contradiction, so e is nonzero.

Case 3. $e < 0$

Having eliminated two possibilities out of a total of three, we naturally arrive at this conclusion. Of course, we also could have reached this conclusion at the outset by noting that

$$e = -\frac{s L_1 C_2}{36 J_3} (3s + 2m) < 0 \tag{6.230}$$

The presentation above serves, once again, the purpose of demonstrating alternate way(s) to deal with a given problem and the diversity in devising a problem-solving strategy when facing more difficult situations.

Let us consider $y' = 0$, which corresponds to $N = 0$. This happens when

$$z_0 = \frac{-c + (c^2 - mce)^{1/2}}{e} \tag{6.231}$$

Additionally, $N < 0$ when $z < z_0$, whereas $N > 0$ when $z > z_0$. Therefore, y has a minimum at z_0.

It is interesting to further explore the expression in parentheses in (6.231). It is

$$Q^2 = c(c - me) = p^2 s^2 (s + m)(s + 2m) \tag{6.232}$$

where

$$p = \frac{s L_1 C_2}{36 J_3} \tag{6.233}$$

Note that

$$Q > 0 \quad \text{and} \quad ps(s + m) < Q < ps(s + 2m) \tag{6.234}$$

Now let us look at y itself. We can rewrite y as

$$y = \frac{h}{g} \tag{6.235}$$

Since $g > 0$, we know that y is positive, negative, or zero when h is so respectively. When will $y = 0$ be the case? This is the case when $h = 0$ and is at

$$z_1 = \frac{-c}{e} = \frac{s^2}{3s + 2m} \tag{6.236}$$

Thus, z_1 is a critical point in that $y < 0$ when $z < z_1$, whereas $y > 0$ when $z > z_1$.

6.6. UNIFORM LOAD ON AN EXTERIOR SPAN OF A THREE-SPAN CONTINUOUS BEAM: GENERAL CASE

The parametric study of L_i, C_i ($i = 1, 2, 3$) was done in Section 3.2.3. Here we will pursue another study, namely the effects of the location of the loaded portion of the span. We take the reaction R_1, as presented in that section, in the form

$$R_1 = \frac{w(h_1 E_2 + h_2 E_3 + h_3 E_4)}{D} \tag{6.237}$$

where

$$h_1 = Z\left[T - \frac{C_2 L_3^2}{3} \right] \tag{6.238}$$

$$h_2 = \frac{-C_3 L_3 Z}{2} \tag{6.239}$$

$$h_3 = \frac{ZL_3}{6J_3} \tag{6.240}$$

as defined earlier.

Since the location of the loaded portion appears only in h_1, h_2, h_3, and E_2, E_3, E_4, we will consider the function f defined as

$$f = h_1 E_2 + h_2 E_3 + h_3 E_4 \tag{6.241}$$

where the h's can be rewritten in terms of the fundamental entities as

$$h_1 = \frac{L_1 L_3^2 C_2 C_3}{18} > 0 \qquad (6.242)$$

$$h_2 = -\frac{L_1 L_3 C_2 C_3}{12} < 0 \qquad (6.243)$$

$$h_3 = \frac{L_1 C_2 C_3}{36} > 0 \qquad (6.244)$$

The function f can be put in the following form

$$f = bg(a) \qquad (6.245)$$

with

$$g(a) = d_3 a^3 + d_2 a^2 + d_1 a + d_0 \qquad (6.246)$$

as a function of the locator of load being our next target.

The coefficients in the expression for $g(a)$ are defined as

$$d_3 = \frac{L_1 C_2 C_3}{36} > 0 \qquad (6.247)$$

$$d_2 = \frac{L_1 C_2 C_3}{24} (b - 2L_3) < 0 \qquad (6.248)$$

$$d_1 = \frac{L_1 C_2 C_3}{36} (2L_3^2 - 3L_3 b + b^2) \qquad (6.249)$$

$$d_0 = \left(\frac{h_1}{2}\right) b + \left(\frac{h_2}{3}\right) b^2 + \left(\frac{h_3}{4}\right) b^3 \qquad (6.250)$$

We have

$$g' = 3d_3 a^2 + 2d_2 a + d_1 \qquad (6.251)$$

where d_3 and d_2 are positive or negative respectively by noting (6.247) and (6.248), but the coefficient d_1 may be positive, negative, or zero. We will explore this now.

For convenience, let

$$K = 2L_3^2 - 3L_3 b + b^2 \qquad (6.252)$$

We see from (6.249) that d_1 and K have the same sign and that $d_1 = 0$ whenever $K = 0$.

Now $K = 0$ when

$$(L_3 - b)(2L_3 - b) = 0 \qquad (6.253)$$

This means $L_3 = b$ or $2L_3 = b$ theoretically. However, $2L_3 = b$ is physically not possible, so only $L_3 - b$ is admissible.

Next, $K < 0$ when $(L_3 - b)$ and $(2L_3 - b)$ have opposite signs. However, this is physically impossible.

Finally, $K > 0$ when

$$(L_3 - b)(2L_3 - b) > 0 \qquad (6.254)$$

This is always the case, with the exception of $L_3 = b$, which corresponds to $K = 0$.

Let us proceed to d_0. We find that a good-looking expression for d_0 is

$$d_0 = \left(\frac{L_1 C_2 C_3}{144} \right) Mb \qquad (6.255)$$

where

$$M = [4L_3^2 - 4L_3 b + b^2] \qquad (6.256)$$

is extremely nice and simple. It is a quadratic expression in either b or L_3 for one thing, and is even a complete squire for another. Note that $d_0 = 0$ whenever $M = 0$ and that d_0 and M have the same sign.

Now $M = 0$ when $(2L_3 - b) = 0$, but this is not possible on grounds of physical meaning. It can be seen clearly that $M > 0$. Thus, $d_0 > 0$. This completes the description of the coefficients in g.

Now let us return to g' first. $g' = 0$ when

$$3a^2 + 3Na + K = 0 \qquad (6.257)$$

where

$$N = b - 2L_3 < 0 \qquad (6.258)$$

It is interesting to note that $N^2 = M$.

The roots of (6.257) are a_1 and a_2 and they have the appearance

$$a_1 = \frac{-3N + Q^{\frac{1}{2}}}{6} \qquad (6.259)$$

$$a_2 = \frac{-3N - Q^{\frac{1}{2}}}{6} \qquad (6.259A)$$

with

$$Q = 3(4L_3^2 - b^2) > 0 \qquad (6.260)$$

and $a_1 < a_2$. Both of the roots shown in (6.259) and (6.259A) are positive real and both are admissible.

Next, we note that

$$g'' = 6d_3a + 2d_2 \qquad (6.261)$$

and is negative at a_1 and positive at a_2. Hence, g has a maximum at a_1 and a minimum at a_2.

Consequently, f and R_1 have a maximum at a_1 and a minimum at a_2.

6.7. TRIANGULAR LOAD ON AN EXTERIOR SPAN OF A THREE-SPAN CONTINUOUS BEAM: GENERAL CASE

The usual approach established earlier will be followed in the parametric study below. First of all, we will review the entities involved in the expressions for the reactions when considering the effects of a particular parameter on the reactions. Then we will express each of the entities under study as a function of the parameter. Utilizing the appropriate method, we will evaluate and draw conclusions from results of this mathematical analysis.

6.7.1. Parametric Study of J_1

In order to maintain uniformity of notations as much as possible, we denote the parameter to be studied as z, while observing that the entities Q, E_i, and A are independent of z and are thus constants in the process of mathematical analysis. We would like to study the effects of z on the reaction R_1. There are several entities that appear in the expression for R_1. Therefore, we will, first of all, deal with these entities and treat them as functions of z.

With the above brief introduction, we have

$$G = h_0 + h_1 z^{-1} \qquad (6.262)$$

where

$$h_0 = \frac{L_1^2 C_2}{3}, \qquad h_1 = \frac{L_1^3}{3} \qquad (6.263)$$

The denominator in the expressions for all reactions is given by

$$D = d_0 + d_1 z^{-1} \qquad (6.264)$$

where

$$d_0 = \left(\frac{L_1 L_3}{6}\right)^2 (3C_2^2 + 4C_2 C_3) \qquad (6.265)$$

$$d_1 = 4\left(\frac{L_1 L_3}{6}\right)^2 L_1 (C_2 + C_3) \tag{6.266}$$

Other entities can be expressed as follows. For D_1, we have the formula

$$D_1 = a_0 + \frac{a_1}{z} \tag{6.267}$$

with

$$a_0 = L_3\left(Q^2 - \frac{C_2 h_0}{3}\right) \tag{6.268}$$

$$a_1 = \frac{-L_3 C_2 h_1}{3} \tag{6.269}$$

For D_2, we have the formula

$$D_2 = b_0 + \frac{b_1}{z} \tag{6.270}$$

with

$$b_0 = \frac{\dfrac{-C_3}{2}}{h_0} \tag{6.271}$$

$$b_1 = \frac{\dfrac{-C_3}{2}}{h_1} \tag{6.272}$$

Finally, for D_3, we have the formula

$$D_3 = c_0 + \frac{c_1}{z} \tag{6.273}$$

with

$$c_0 = \frac{h_0}{6 J_3} \tag{6.274}$$

$$c_1 = \frac{h_1}{6 J_3} \tag{6.275}$$

As a result of the above introduction, we have

$$R_1 = \frac{C}{D} \tag{6.276}$$

where C is a constant as far as z is concerned.

For convenience in subsequent development, we will let

$$y = \frac{1}{d_0 + \dfrac{d_1}{z}} \tag{6.277}$$

Thus,

$$R_1 = cy \tag{6.277A}$$

Since $d_1 > 0$, we have

$$y' = d_1 \left[z \left(d_0 + \frac{d_1}{z} \right) \right]^2 > 0 \tag{6.278}$$

Therefore, y is an increasing function of z.

Note that the function y can be rewritten as

$$y = \frac{z}{d_0 z + d_1} = \frac{\left(1 - \dfrac{\dfrac{d_1}{d_0}}{z + \dfrac{d_1}{d_0}} \right)}{d_0} \tag{6.279}$$

which is always positive.

Next, let us proceed to the investigation of R_4. Let

$$N = g_1 E_2 + g_2 E_3 + g_3 E_4 + g_4 E_5 \tag{6.280}$$

Then, using the results presented above for the entities D_i and the definitions of g_i, we have

$$N = n_0 + \frac{n_1}{z} \tag{6.281}$$

where

$$n_0 = a_0 e_1 + b_0 e_2 + c_0 e_3 \tag{6.282}$$

$$n_1 = a_1 e_1 + b_1 e_2 + c_1 e_3 \tag{6.283}$$

with e_i defined as

$$e_1 = \left(\frac{w}{b}\right)(-aE_2 + E_3), \qquad e_2 = \left(\frac{w}{b}\right)(-aE_3 + E_4),$$

$$e_3 = \left(\frac{w}{b}\right)(-aE_4 + E_5) \tag{6.284}$$

Thus, we obtain

$$R_4 = \frac{N}{D} = \left(\frac{1}{d_0}\right)\left[n_0 + \frac{P}{z + r}\right] \tag{6.285}$$

where

$$r = \frac{d_1}{d_0} > 0 \tag{6.286}$$

Again for convenience in subsequent work, we define a new function y as

$$y = \frac{P}{z + r} \tag{6.287}$$

where

$$P = n_1 - \frac{n_0 d_1}{d_0} \tag{6.288}$$

Thus,

$$y' = -P(z + r)^{-2} \tag{6.289}$$

We see that $y' > 0$ if $P < 0$, whereas $y' < 0$ where $P > 0$. Moreover, $y' = 0$ if $P = 0$.

It is clear from these statements that P is of fundamental importance. On reviewing the elements that make up P, we know that a thorough examination of these elements will prove to be useful and interesting as well, so let us start with n_0 and n_1.

Simple algebraic manipulation shows that $a_0 < 0$, $a_1 < 0$, $b_0 < 0$, $b_1 < 0$, $c_0 > 0$, and $c_1 > 0$. It has been proved (see Chapter 1) that

$$E_3 - aE_2 > 0, \qquad E_4 - aE_3 > 0, \qquad E_5 - aE_4 > 0 \tag{6.290}$$

Therefore,

$$e_i > 0 \quad \text{for } i = 1, 2, 3 \tag{6.291}$$

As a result, it is possible that both n_0 and n_1 may be positive, negative, or zero (not necessarily at the same time, of course).

Next, we have the coefficients in D, namely d_0 and d_1, to be reviewed. Both of them are positive, and both have to do with C_2, C_3, L_1, and L_3 and are increasing functions of these fundamental entities.

It is very interesting and important to note that the ratio between these two coefficients in D plays a crucial role in the behavior of P and thus of y and R_4, as will be seen a little later.

Now let us return to the determining component P in both y and y'. On the surface, we see that when $P = 0$, we have $y = 0$ and $y' = 0$. The former means that R_4 is a constant n_0/d_0 regardless of the value z. The latter signifies the value of y when $P = 0$ is a stationary value.

What about the cases where $P > 0$ and $P < 0$? $P > 0$ means

$$n_1 > n_0 r \tag{6.292}$$

or

$$\frac{n_1}{n_0} > r \quad \text{for } n_0 > 0 \tag{6.292A}$$

The case $P < 0$ holds if and only if

$$n_1 < n_0 r \tag{6.293}$$

or

$$\frac{n_1}{n_0} < r \quad \text{for } n_0 > 0 \tag{6.293A}$$

and $P = 0$ holds when

$$n_1 = n_0 r \tag{6.294}$$

or

$$\frac{n_1}{n_0} = r \quad \text{for } n_0 > 0 \tag{6.294A}$$

Note that Formulae (6.292A), (6.293), and (6.294A) all state the significance of the relative magnitude of two ratios. One is between the coefficients in N, whereas the other is between those in D. Moreover, there is an interesting lower bound L_1/C_2 for r, which is expressed as follows:

$$r > \frac{L_1}{C_2} \tag{6.295}$$

Note also that, from (6.292A) and (6.294A), for $P > 0$ or $P = 0$ to hold, it is necessary that $n_1/n_0 > 0$ or $= 0$ respectively, which means that n_1 and n_0 must be of the same sign, where n_0 is nonzero. Also, only when $P < 0$ is $n_1/n_0 < 0$ possible, because $r > 0$ and the very condition that specifies $P < 0$. These remarks apply when n_0 is nonzero. In this connection here, if $n_0 = 0$, then we simply have

$$P = n_1 \tag{6.296}$$

As a result, we have $P > 0$, < 0, or $= 0$ when $n_1 > 0$, < 0, or $= 0$ respectively. This naturally leads to the question: What are the consequences if n is zero? We then have

$$N = \frac{n_1}{z}, \qquad R_4 = \frac{\left(\dfrac{1}{d_0}\right)n_1}{z + r} \tag{6.297}$$

What if both n_0 and n_1 are zero? Well, we have $N = 0$, and D is always positive. Therefore, it is plausible that $R_4 = 0$ theoretically. But is it possible, mathematically, that both n_1 and n_0 are zero at the same time?

That is an interesting question which will invoke an even more interesting investigation process and eventual answer. It can be left as a future endeavor for the reader who chooses to skip the next several paragraphs under the heading "The Black Hole—Or Is It?" This reader probably will get more than he or she bargained for. Additionally, as was indicated earlier, it is not the intention of this book to encompass answers to all reasonable questions. Rather, the purpose is to show the philosophy, methodology, and approaches, as well as possibility of the existence of new frontiers and to inspire the reader to do the exploration on his or her own so that applying this book will be fun and useful.

The Black Hole—Or Is It?

We begin by looking at the definitions of n_0 and n_1. To do this, we need to display the definitions of the symbols and the attributes of these concepts defined therein. The list is as follows:

$$a_0 = \frac{-C_2^2 L_2^2 L_3}{12}, \qquad b_0 = \frac{-C_2 C_3 L_1^2}{6}, \qquad c_0 = \frac{C_2 L_1^2}{18 J_3} \tag{6.298}$$

$$a_1 = \frac{-C_2 L_3 L_1^3}{9}, \qquad b_1 = \frac{-C_3 L_1^3}{6}, \qquad c_1 = \frac{L_1^3}{18 J_3} \tag{6.299}$$

After some algebraic manipulations and simplification, we arrive at

$$n_0 = \left(\frac{C_2 L_1^2}{36}\right) M \tag{6.300}$$

with

$$M = -3C_2L_3e_1 - 6C_3e_2 + \frac{2e_3}{J_3} \qquad (6.301)$$

and

$$n_1 = \left(\frac{L_1^3}{36}\right)N \qquad (6.302)$$

where

$$N = -4C_2L_3e_1 - 6C_3e_2 + \frac{2e_3}{J_3} \qquad (6.303)$$

Pay close attention to the appearance of M and N, because they look very much like twin brothers. In fact, the only difference is the numerical value of the first coefficient. In precise terms,

$$N = M - C_2L_3e_1 \qquad (6.304)$$

Now let us return to our original question: Can both n_0 and n_1 be zero? For n_0 to be zero, we must have $M = 0$ because the other factor of n_0 is positive. From this, we have

$$3C_2L_3e_1 = \frac{2e_3}{J_3} - 6C_3e_2 \qquad (6.305)$$

For n_1 to be zero, $N = 0$ is required, because once again the other factor of n_1 is positive (and of course nonzero). As a result of this, we have

$$4C_2L_3e_1 = \frac{2e_3}{J_3} - 6C_3e_2 \qquad (6.306)$$

Looking at the right side of both equalities (6.305) and (6.306) first, we see that they are exactly the same. Having done that, let us look at the left side of these formulae again. Surprise and surprise—N and M are not perfect twins. Thus, the principle is that n_0 and n_1 cannot be zero at the same time. That solves the mystery we had before digging in the black hole.

Since we are already in, we might just as well go one step further and dig a little deeper. We have seen the ratio n_1/n_0 on several occasions where n_0 is nonzero and are naturally very curious about its attributes. Let us call this ratio t. Thus

$$t = \frac{n_1}{n_0} \qquad (6.307)$$

which is equal to

$$t = \frac{L_1 N}{C_2 M} \qquad (6.308)$$

Note that both C_2 and L_1 are positive. Hence, t can be positive or negative depending on the sign of N/M. Furthermore, t is zero when N is zero. Let us look at this matter case by case as follows.

Case 1. $t = 0$

This happens when $N = 0$, as noted before. From the definition of N, this means

$$\frac{2e_3}{J_3} = 4C_2 L_3 e_1 + 6C_3 e_2 \qquad (6.309)$$

Another implication of $t = 0$ is

$$M > 0 \qquad (6.310)$$

which is the result of $N = 0$ and $M > N$ [see (6.304) above regarding the relationship between M and N]. Consequently, $n_1 = 0$ and $n_0 > 0$. From this last statement, we have via (6.288), upon invoking the definition of r,

$$P = -n_0 r < 0 \qquad (6.311)$$

where

$$n_0 = \frac{C_2^2 L_1^2 L_3 e_1}{36} \qquad (6.312)$$

From the above, we see that P is directly proportional to r and a very neat, nice-looking n_0, in addition to having a much simpler expression than (6.288) itself. Even though (6.285) is not much simplified in form, actual expressions for both n_0 and P are much simpler than before, and just because of this we have

$$R_1 = \left(\frac{n_0}{d_0}\right)\left(1 - \frac{r}{z + r}\right) \qquad (6.313)$$

The last formula says that R_4 is positive and proportional to the ratio n_0/d_0, which is the absolute upper limit of R_4. Also, from (6.311) and (6.289), we see that $y' > 0$. Therefore, y is an increasing function of z and so is R_4 by virtue of Formulae (6.285) and (6.287).

Case 1 appears to be the story of an innocent-looking pair of "pseudo" twin brothers M and N which are powerful or even magical actors. But they are not finished with their show yet. In fact, we have only seen one scene, namely $t = 0$, and there are two more scenes, $t < 0$ and $t > 0$, to go.

Case 2. t < 0

Again, we look at M and N for clues. This is the case where M and N have opposite signs. Due to the way they are, this automatically puts M on the positive side and N on the other side (i.e., $M > 0$ and $N < 0$). This means, then, that $n_0 > 0$ and $n_1 < 0$. From the last statement and Formula (6.288), we see immediately that $P < 0$. As a result of this and from (6.289), we have $y' > 0$. Thus, y is an increasing function of z and so is R_4 [see (6.285)].

Now we ask: Is it possible for R_4 to be zero or negative? The answer is yes. From (6.285) and (6.288), $R_4 = 0$ requires that

$$z = \frac{-n_1}{n_0} \tag{6.314}$$

which is possible because $t = n_1/n_0$ is negative in the case under study here.

Of course, the condition for $R_4 > 0$ is, from (6.285) and because $d_0 > 0$, simply

$$n_0 + \frac{P}{z + r} > 0 \tag{6.315}$$

which leads, with the aid of (6.288), to $z > -n_1/n_0$. Similarly, the condition $R_4 < 0$ will lead to $z < -n_1/n_0$, which is possible for any given n_1, n_0. Thus, we see that R_4 may be positive, negative, or zero depending on the relative magnitude of z and t specified above.

Case 3. t > 0

Since $t = L_1 N/(C_2 M)$ and both L_1 and C_2 are positive, it follow that the sign of t agrees with that of N/M. This is the present case when M and N have the same sign. Thus, there are two possibilities for this to hold. One is when both M and N are positive, and the other is when M and N are negative at the same time. We will review these two sources of M and N that have the same sign next.

(a) Both M and N are positive. We note that $N/M < 1$. Thus,

$$t < \frac{L_1}{C_2} \tag{6.316}$$

That is, t has an upper bound $L_1/C_2 > 0$.

It is obvious that the cases where $t < 0$ and $t = 0$ both satisfy (6.316). So far, no t can be greater than L_1/C_2. Let us check the possibilities for $P = 0$, < 0, or > 0.

First, $P = 0$ means, from (6.288) and the definition of t, that $t = r$. However, we know from (6.293) that $r > L_1/C_2$, so $P = 0$ would bring us to $t > L_1/C_2$, which is not possible in view of (6.316). Therefore, we reach the conclusion that t cannot be equal to r, as well as the fact that P is nonzero.

If $P > 0$, then from (6.288) and (6.307) we have $t > r$, which means that $t > L_1/C_2$ from the property of r, namely $r > L_1/C_2$. But $t > L_1/C_2$ is not possible, given the validity of (6.316). Therefore, P is not positive, so P must be negative, and this easily can be confirmed by other methods also.

From (6.285), we see that R_4 has an upper bound $n_0/d_0 > 0$ since n_0 is positive as a result of M, C_2, and L_1 being positive. Can R_4 be zero or negative then? Well, let us see.

For R_4 to be zero, we need

$$zn_0 = -P - rn_0 \quad \text{and} \quad z > 0 \tag{6.317}$$

Thus, $-P - rn_0 > 0$, but from (6.288) we have $P = n_0(t - r)$. Therefore, we arrive at $-n_0 t > 0$. This means that $-n_0 > 0$ is required. However, it is given that both M and N are positive, and so is, therefore, n_0. Thus, the supposition that R_4 can be zero is not valid. Similarly, if we assume that $R_4 < 0$, then we will reach the logical conclusion that $n_0 z < -P - n_0 r$. This means that $n_0 z < -n_0 t < 0$. As a result of this, and $n_0 > 0$ and $t > 0$, we have $z < t < 0$, which is not admissible since z is positive.

Based on the above, we conclude that R_4 cannot be negative either under the conditions stipulated above, including both M and N being positive. In other words, R_4 must be positive.

(b) Both M and N are negative. We observe that, in this case, $N/M > 0$ and (6.316) still holds, along with the statement that L_1/C_2 is an upper bound for t. Also, we have the same conclusion as in (a): R_4 can be positive only as proved by several different approaches. However, there is one big difference between (a) and (b): n_0/d_0 is a lower bound for R_4 because here in (b), $P > 0$. If we want to see why $P > 0$, we simply note that $P = n_0(t - r)$ and the fact that here (1) $(t - r) < 0$ because $r > L_1/C_2 > t$ and (2) $n_0 < 0$ because $M < 0$ is given, and n_0 is just M multiplied by a positive entity $C_2 L_1^2/36$. This concludes the description of the black hole.

6.7.2. Parametric Study of J_2

Let us denote the parameter J_2 by z. Here, entities entering the expressions for R_1 and R_4 have the following features:

$$Q = \frac{L_1 L_2}{6z} \tag{6.318}$$

A_1, A_2, and A_3 are independent of z.

$$G = \left[C_1 + \frac{L_2}{z}\right]\frac{L_1^2}{3} \tag{6.319}$$

$$D_1 = \frac{a_1}{z} + \frac{a_2}{z^2}, \qquad a_1 = \frac{-C_1 L_1^2 L_2 L_3}{9}, \qquad a_2 = \frac{-L_1^2 L_2^3}{12} \tag{6.320}$$

$$D_2 = b_0 + \left(\frac{b_1}{z}\right), \qquad b_0 = \frac{-C_1 C_3 L_1^2}{6}, \qquad b_1 = \frac{-C_3 L_1^2 L_2}{6} \tag{6.321}$$

$$D_3 = c_0 + \frac{c_1}{z}, \qquad c_0 = \frac{C_1 L_1^2}{18 J_3}, \qquad c_1 = \frac{L_1^2 L_2}{18 J_3} \tag{6.322}$$

$$D = d_2 z^{-2} + d_1 z^{-1} + d_0 \tag{6.323}$$

where

$$d_2 = \frac{(L_1 L_2 L_3)^2}{2} \tag{6.324}$$

$$d_1 = 4L_2 \left[\frac{L_1 L_3}{6}\right]^2 (C_1 + C_3) \tag{6.325}$$

$$d_0 = 4C_1 C_3 \left(\frac{L_1 L_3}{6}\right)^2 \tag{6.326}$$

With this introduction of the necessary notations, we begin our study of R_1 as follows.

$$R_1 = \frac{CQ}{D} \tag{6.327}$$

where

$$C = f_1 E_2 + f_2 E_3 + f_3 E_4 + f_4 E_5 \tag{6.328}$$

is independent of z.

Let us define y by

$$y = \frac{Q}{D} \tag{6.329}$$

Then

$$y = \frac{L_1 L_2}{6} \left(\frac{z}{H}\right) \tag{6.330}$$

where

$$H = (d_2 + d_1 z + d_0 z^2) \tag{6.331}$$

is positive.

From (6.330), we see that y is positive also except at

$$z = 0, \text{ meaning } y = 0 \tag{6.332}$$

the impractical case of $J_2 = 0$.

Now

$$y' = \frac{d_2 - d_0 z^2}{H^2} \tag{6.333}$$

Hence,

$$y' = 0 \text{ when } d_2 - d_0 z^2 = 0 \tag{6.334}$$

This means that

$$z_0 = \left(\frac{d_2}{d_0}\right)^{1/2} = \frac{3^{1/2} L_2}{2(C_1 C_3)^{1/2}} \tag{6.335}$$

Moreover,

$$y' > 0 \text{ when } z < z_0, \qquad y' < 0 \text{ when } z > z_0 \tag{6.336}$$

Hence, y has a maximum at z_0.

We now proceed to look at R_4. Let us introduce a convenient notation N defined by

$$N = g_1 E_2 + g_2 E_3 + g_3 E_4 + g_4 E_5 \tag{6.337}$$

which is equal to

$$N = D_1 e_1 + D_2 e_2 + D_3 e_3 \tag{6.338}$$

where e_i ($i = 1, 2, 3$) is as defined in Section 6.7.1.

Let us review the contents of D_1, D_2, and D_3 in some detail. Regarding D_1, both a_1 and a_2 are negative. For D_2, both b_0 and b_1 are also negative. As to D_3, both c_0 and c_1 are positive. These observations enable us to assert that both D_1 and D_2 are negative, whereas D_3 is positive. Moreover, all the e_i are positive. As a result, we see that N may be positive, negative, or zero.

We now rearrange the terms of N according to the powers of z as follows.

$$N = n_2 z^{-2} + n_1 z^{-1} + n_0 \tag{6.339}$$

where

$$n_2 = a_2 e_1, \qquad n_1 = a_1 e_1 + b_1 e_2 + c_1 e_3, \qquad n_0 = b_0 e_2 + c_0 e_3 \tag{6.340}$$

We can rewrite n_0 as a product of two factors, one of which contains e_2, e_3, and L_3 and the other which does not. We call the former factor K and the latter H for convenience. Thus,

$$n_0 = KH \tag{6.341}$$

where

$$K = e_3 - 3e_2L_3 \tag{6.342}$$

$$H = \frac{C_1L_1^2}{18J_3} \tag{6.343}$$

Noting that e_2, e_3, and L_3 are all positive in the expression for K, and that n_0 is directly proportional to K and H, with H being positive, we immediately come to the following important conclusion: n_0 is positive, negative, or zero according to whether K is positive, negative, or zero respectively. Now

$$K > 0 \text{ when } \frac{e_3}{3e_2} > L_3 \tag{6.344}$$

$$K < 0 \text{ when } \frac{e_3}{3e_2} < L_3 \tag{6.345}$$

$$K = 0 \text{ if and only if } \frac{e_3}{3e_2} = L_3 \tag{6.346}$$

Formulae (6.344–6.346) have this message for us: There is a very special span length L_3, called the critical span length, which determines the fate of n_0, the most influential coefficient in N, as will be seen below.

The expression for R_4 is

$$R_4 = \frac{N}{D} \tag{6.347}$$

which can be rewritten as

$$R_4 = \frac{n_0z^2 + n_1z + n_2}{d_0z^2 + d_1z + d_2} \tag{6.348}$$

The sign of R_4 is determined completely by that of the numerator, henceforth called M for convenience, in (6.348), because the denominator, called D_0, is equal to D, which is positive, multiplied by z^2. Moreover, R_4 is zero when M is.

Now we see that the leading coefficient in the numerator in (6.348) is the "promising" n_0 that was mentioned before.

Of special interest is the case where n_0 is zero. Here, the original

$$M = n_0z^2 + n_1z + n_2 \tag{6.349}$$

becomes simply

$$M = n_1z + n_2 \tag{6.350}$$

If we set $M = 0$ in (6.350), we have

$$z = \frac{-n_2}{n_1} \tag{6.351}$$

which is possible provided n_1 and n_2 have opposite signs. We know that n_2 is negative because a_2 is negative, e_1 is positive, and n_2 is the product of these two entities. This leads to the consideration of n_1, which must be positive to make (6.351) physically meaningful.

Next, let us look at the general case where n_0 is nonzero. If we set $M = 0$, then the roots of this equation are

$$z_1 = \frac{-n_1 + S^{1/2}}{2n_0} \tag{6.352}$$

$$z_2 = \frac{-n_1 - S^{1/2}}{2n_0} \tag{6.353}$$

where

$$S = n_1 - 4n_0n_2 \tag{6.354}$$

For real roots, it is required that

$$S > 0 \quad \text{or} \quad S = 0 \tag{6.355}$$

With (6.355) and the condition $n_0 < 0$, we have two cases to consider:

1. $n_1 < 0$ implies that there are no positive roots.
2. $n_1 > 0$ means that there are two positive roots, as given in (6.352) and (6.353).

With (6.355) and the condition $n_0 > 0$, we also have two cases to consider:

1. $n_1 < 0$ signifies that there is only one positive root, given by (6.352).
2. $n_1 > 0$ leads to the same conclusion as for the case just described.

The reaction R_4 can be positive, negative, or zero, as mentioned earlier. Thus,

$$R_4 > 0 \quad \text{when} \quad n_0(z - z_1)(z - z_2) > 0 \tag{6.356}$$

$$R_4 < 0 \quad \text{when} \quad n_0(z - z_1)(z - z_2) < 0 \tag{6.357}$$

$$R_4 = 0 \quad \text{when} \quad n_0(z - z_1)(z - z_2) = 0 \tag{6.357A}$$

Note that in (6.356–6.357A) the three factors n_0, $(z - z_1)$, and $(z - z_2)$ are important and must all be considered because they all contribute to the final results.

Next, we will examine the first derivative of R_4. Once again, for the sake of keeping with the trend in this section and at the same time changing pace from the rest of the book, we will do $n_0 = 0$ first.

For the very interesting case where $n_0 = 0$, we have

$$R_4 = \frac{P}{D} \tag{6.358}$$

where

$$P = p_2 z^2 + p_1 z + p_0 \tag{6.359}$$

with

$$p_2 = -d_0 n_1, \qquad p_1 = -2d_0 n_2, \qquad p_0 = d_2 n_1 - d_1 n_2 \tag{6.360}$$

Now $P = 0$ when

$$z = \frac{-p_1 + T^{\frac{1}{2}}}{2p_2} \tag{6.361}$$

or

$$z = \frac{-p_2 - T^{\frac{1}{2}}}{2p_2} \tag{6.361A}$$

where

$$T = p_1^2 - 4p_0 p_2 \tag{6.362}$$

For real roots of $P = 0$, it is required to have

$$T > 0 \quad \text{or} \quad T = 0 \tag{6.363}$$

When (6.363) is satisfied, we then check for positive roots as admissible ones. When $P = 0$, we check for a possible maximum or minimum of R_4. When $P > 0$, we know that R_4 is an increasing function of z. When $P < 0$, then R_4 is a decreasing function z.

For the general case where n_0 is nonzero, we have

$$R_4 = \left(\frac{1}{d_0}\right)\left[n_0 + \frac{P}{\dfrac{D_0}{d_0}}\right] \tag{6.364}$$

where

$$P = p_1 z + p_0 \tag{6.365}$$

with

$$p_1 = n_1 - \frac{n_0 d_1}{d_0}, \qquad p_0 = n_2 - \frac{n_0 d_2}{d_0} \qquad (6.366)$$

The first derivative of R_4 with respect to z is

$$R_4' = \left(\frac{1}{d_0}\right)\left[\frac{r_2 z^2 + r_1 z + r_0}{\left(\dfrac{D}{d}\right)^2}\right] \qquad (6.367)$$

where

$$r_2 = \frac{n_0 d_1}{d_0} - n_1, \qquad r_1 = \left(\frac{n_0 d_2}{d_0} - n_2\right), \qquad r_0 = \frac{n_1 d_2 - n_2 d_1}{d_0} \qquad (6.368)$$

It is interesting to note that $r_2 = -p_1$ and $r_1 = -2p_0$.

$R_4' = 0$ when the numerator of R_4' is zero. This is when

$$z_1 = \frac{-r_1 + (r_1^2 - 4r_0 r_2)^{1/2}}{2r_2} \qquad (6.369)$$

or

$$z_2 = \frac{-r_1 - (r_1^2 - 4r_0 r_2)^{1/2}}{2r_2} \qquad (6.370)$$

Again, we check the real positive root(s) requirements for admissibility.

Finally, we have

$$R_4' > 0 \quad \text{when} \quad r_2(z - z_1)(z - z_2) > 0 \qquad (6.371)$$

$$R_4' < 0 \quad \text{when} \quad r_2(z - z_1)(z - z_2) < 0 \qquad (6.372)$$

The above is for nonzero r_2. If $r_2 = 0$, then $R' = 0$ requires, for physical meaning,

$$z = \frac{-r_0}{r_1} > 0 \qquad (6.372A)$$

6.7.3. Parametric Study of J_3

We will replace all J_3 with z and examine the entities involved in the expressions for R_1 and R_4.

First, we observe that Q, G, and E are independent of z, and we denote the A_i as

$$A_1 = \frac{a_1}{z}, \qquad A_2 = \frac{a_2}{z}, \qquad A_3 = \frac{a_3}{z} \tag{6.373}$$

with

$$a_1 = \frac{L_3^3}{3}, \qquad a_2 = \frac{-L_3^2}{2}, \qquad a_3 = \frac{L_3}{6} \tag{6.374}$$

Next, we note that

$$D_1 = L_1^2 L_3 \left(\frac{C_2^2}{12} + \frac{C_1 C_2}{9} \right) \tag{6.375}$$

is independent of z, in contrast to

$$D_2 = \frac{-L_3 G}{2z}, \qquad D_3 = \frac{G}{6z} \tag{6.376}$$

which are functions of z.

Finally, the denominator in the expressions for R_1 and R_4 is

$$D = d_0 + \frac{d_1}{z} \tag{6.377}$$

where

$$d_0 = \left[\frac{L_1 L_3}{6} \right]^2 (3C_2^2 + 4C_1 C_2) \tag{6.378}$$

$$d_1 = 4 \left[\frac{L_1 L_3}{6} \right]^2 L_3 (C_1 + C_2) \tag{6.379}$$

Thus, we obtain

$$R_1 = \frac{QN}{D} \tag{6.380}$$

where

$$N = f_1 E_2 + f_2 E_3 + f_3 E_4 + f_4 E_5 = \frac{a_1 e_1 + a_2 e_2 + a_3 e_3}{z} = \frac{C}{z} \tag{6.381}$$

with

$$C = (a_1 e_1 + a_2 e_2 + a_3 e_3) \tag{6.382}$$

as a constant relative to z.

Let

$$y = \frac{1}{d_0 z + d_1} \tag{6.383}$$

Then

$$y' = -d_0(d_0 z + d_1)^{-2} < 0 \tag{6.384}$$

Therefore, y is a decreasing function of z.

Now let us move to the evaluation of R_4. The numerator in the expression for R_4 is

$$g_1 E_2 + g_2 E_3 + g_3 E_4 + g_4 E_5 = n_0 + \frac{n_1}{z} \tag{6.385}$$

where

$$n_0 = D_1 e_1, \qquad n_1 = \left(\frac{-L_3 G}{2}\right) e_2 + \left(\frac{G}{6}\right) e_3 \tag{6.386}$$

Thus

$$R_4 = \frac{n_0 + \dfrac{n_1}{z}}{d_0 + \dfrac{d_1}{z}} = \left(\frac{1}{d_0}\right) \left[n_0 + \frac{P}{z + \dfrac{d_1}{d_0}} \right] \tag{6.387}$$

where

$$P = n_1 - \frac{n_0 d_1}{d_0} \tag{6.388}$$

From (6.388), it is clear that

$$P = 0 \text{ when } n_1 = \frac{n_0 d_1}{d_0} \tag{6.389}$$

$$P < 0 \text{ when } n_1 < \frac{n_0 d_1}{d_0} \tag{6.390}$$

$$P > 0 \text{ when } n_1 > \frac{n_0 d_1}{d_0} \tag{6.391}$$

We see that the last four formulae are our old friends from Section 6.7.1.

Finally, we have

$$R_4' = \frac{-P}{d_0}\left(z + \frac{d_1}{d_0}\right)^{-2}$$ (6.392)

which is zero, negative, or positive when P is zero, positive, or negative respectively.

6.7.4. Parametric Study of L_1

The independent variable is again denoted by z. The entities A_1, A_2, and A_3 are free from z and thus are constants in this investigation. However, Q, G, D_1, D_2, D_3, and D are functions of z. As a result, we have

$$R_1 = \frac{\dfrac{CC_2}{6}}{z(d_0 + d_1 z)}$$ (6.393)

where

$$C = f_1 E_2 + f_2 E_3 + f_3 E_4 + f_4 E_5$$ (6.394)

is a constant. The entities d_0 and d_1 are defined as

$$d_0 = \left(\frac{L_3}{6}\right)^2 (3C_2^2 + 4C_2 C_3), \qquad d_1 = \left[\frac{L_3^2}{9J_1}\right](C_2 + C_3)$$ (6.395)

Let

$$y = \frac{1}{z(d_0 + d_1 z)}$$ (6.396)

Then

$$R_1 = \left(\frac{CC_2}{6}\right)y$$ (6.393A)

and $y' < 0$ for all z. Therefore, y as well as R_1 are decreasing functions of z.

Regarding the reaction R_4, we have

$$R_4 = \frac{N}{D}$$ (6.397)

where

$$N = z_2(n_1 z + n_0)$$ (6.398)

with

$$n_1 = a_3 e_1 - \left(\frac{C_3}{2}\right) h_3 e_2 + \left[\frac{1}{6J_3} h_3 e_3\right] \tag{6.399}$$

$$n_0 = a_2 e_1 - \left(\frac{C_3}{2}\right) h_2 e_2 + \left[\frac{1}{6J_3} h_2 e_3\right] \tag{6.400}$$

$$h_2 = \frac{C_2}{3}, \qquad h_3 = \frac{1}{3J_1} \tag{6.401}$$

Thus,

$$R_4 = \frac{n_0 + n_1 z}{d_0 + d_1 z} \tag{6.402}$$

which can be rewritten as

$$R_4 = \left(\frac{1}{d_1}\right) \left[n_1 + \frac{P}{z + \dfrac{d_0}{d_1}}\right] \tag{6.403}$$

where

$$P = n_0 - \frac{n_1 d_0}{d_1} \tag{6.404}$$

We see some familiar expressions as follows:

$$P < 0 \text{ when } n_0 - \frac{n_1 d_0}{d_1} < 0 \tag{6.405}$$

$$P = 0 \text{ when } n_0 - \frac{n_1 d_0}{d_1} = 0 \tag{6.406}$$

$$P > 0 \text{ when } n_0 - \frac{n_1 d_0}{d_1} > 0 \tag{6.407}$$

Also,

$$R_4' = \frac{-P}{d_1} \left[z + \frac{d_0}{d_1}\right]^{-2} \tag{6.408}$$

Hence, R_4 is a decreasing or increasing function of z for $P > 0$ or $P < 0$ respectively. Moreover, the rate of change is decreasing also. Note that both n_0 and n_1 can be positive, negative, or zero.

6.7.5. Parametric Study of L_2

Again, we denote the independent variable by z. Then

$$R_1 = \frac{CQ}{D} \tag{6.409}$$

where

$$C = f_1E_2 + f_2E_3 + f_3E_4 + f_4E_5 \tag{6.410}$$

is a constant and

$$Q = \left(\frac{L_1}{6J_2}\right)z \tag{6.411}$$

$$D = d_2z^2 + d_1z + d_0 \tag{6.412}$$

with

$$d_2 = 3\left[\frac{\frac{L_1L_3}{6}}{J_2}\right]^2, \qquad d_1 = 4\left[\frac{L_1L_3}{6}\right]^2\frac{C_1 + C_3}{J_2}, \tag{6.412A}$$

$$d_0 = 4\left[\frac{L_1L_3}{6}\right]^2(C_1C_3)$$

It is convenient to consider

$$\frac{Q}{D} = \left[\frac{L_1}{6J_2}\right]y \tag{6.413}$$

where

$$y = \frac{z}{D} \tag{6.414}$$

Then

$$y' = \frac{d_0 - d_2z^2}{D^2} \tag{6.415}$$

Note that R_1' and y' have the same sign. $R_1' = 0$ when $y' = 0$, that is, at

$$z = \left(\frac{d_0}{d_2}\right)^{1/2} \tag{6.416}$$

and y has a maximum here and so does R_1.

Regarding the reaction R_4, we have

$$R_4 = \frac{n_2 z^2 + n_1 z + n_0}{d_2 z^2 + d_1 z + d_0} = \left(\frac{1}{d_2}\right)\left[n_2 + \frac{\dfrac{P}{D}}{d_2}\right] \tag{6.417}$$

where

$$n_2 = a_2 e_1 > 0, \qquad n_1 = a_1 e_1 + b_1 e_2 + c_1 e_3,$$

$$n_0 = a_0 e_1 + b_0 e_2 + c_0 e_3 \tag{6.418}$$

Note that n_1 and n_0 may be positive, negative, or zero.

The entity P is defined as

$$P = p_1 z + p_0 \tag{6.419}$$

where

$$p_1 = n_1 - \frac{n_2 d_1}{d_2}, \qquad p_0 = n_0 - \frac{n_2 d_0}{d_2} \tag{6.420}$$

Again, p_1 and p_0 can be positive, negative, or zero.

Taking the first derivative of R_4 with respect to z, we obtain

$$R_4' = \frac{\left(\dfrac{1}{d_2}\right)N}{\left(\dfrac{D}{d_2}\right)^2} \tag{6.421}$$

Thus, R_4' and N have the same sign. $R_4' = 0$ when $N = 0$, where N is defined as

$$N = -p_1 z^2 - 2p_0 z + \left(\frac{p_1 d_0}{d_2} - \frac{p_0 d_1}{d_2}\right) \tag{6.422}$$

The roots of $N = 0$ can be found easily via the quadratic formula. Of course, we will check for the conditions specifying real positive roots. Once the roots are obtained, we can determine the maximum or minimum value(s) of the reaction and the corresponding values of the independent variable in the usual manner as presented earlier. This is a good exercise for the interested reader. Note that $p_1 = 0$ leads to the condition $n_1 = n_2 d_1/d_2$, while $p_0 = 0$ means $n_0 = n_2 d_0/d_2$. We can ask the following questions. First, can both of these conditions be satisfied? If they can, what does that mean? Is it possible that $n_0 = 0$? If so, when will it happen and what will its physical significance be? These are interesting and important questions and are worthwhile to pursue. We will leave them to future endeavors.

6.7.6. Parametric Study of L_3

We denote the parameter as z. The reaction R_1 is

$$R_1 = \frac{CN}{z(d_1z + d_0)} \tag{6.423}$$

where

$$C = \frac{Q}{6J_3} \tag{6.424}$$

$$N = 2e_1z^2 - 3e_2z = e_3 \tag{6.425}$$

R_1 can be rewritten as

$$R_1 = \left[\frac{C}{d_1}\right]\left[2e_1 + \frac{P}{z^2 + f_1z}\right] \tag{6.426}$$

where

$$f_1 = \frac{d_0}{d_1}, \qquad P = e_3 - p_1z \tag{6.427}$$

$$p_1 = 3e_2 + 2e_1f_1 > 0 \tag{6.428}$$

Note that $P < 0$, $P = 0$, or $P > 0$ when $e_3 - p_1z < 0$, $= 0$, or > 0 respectively. That is,

$$P < 0 \text{ when } z > \frac{e_3}{p_1} \tag{6.429}$$

$$P = 0 \text{ when } z = \frac{e_3}{p_1} \tag{6.430}$$

$$P > 0 \text{ when } z < \frac{e_3}{p_1} \tag{6.431}$$

$R_1' < 0$, $= 0$, or > 0 when $T < 0$, $= 0$, or > 0 respectively, where

$$T = p_1z^2 - 2e_3z - f_1e_3 \tag{6.432}$$

$T = 0$ has two roots, but only one is positive and therefore admissible, namely

$$z = \frac{e_3 + S^{1/2}}{p_1} \quad \text{with } S = e_3^2 + p_1f_1e_3 \tag{6.433}$$

Here R_1 has a minimum.

Let us take care of R_4 now.

$$R_4 = \frac{n_1z + n_0}{z^2(d_1z + d_0)} \tag{6.434}$$

where

$$n_1 = a_1 e_1 + b_1 e_2 < 0, \qquad n_0 = D_3 e_3 > 0 \qquad (6.435)$$

with

$$a_1 = Q^2 - \frac{C_2 G}{3} < 0, \qquad b_1 = \frac{-G}{2J} < 0 \qquad (6.436)$$

$R_4' < 0, = 0,$ or > 0 when $U > 0, = 0,$ or > 0 respectively, where

$$U = u_2 z^2 + u_1 z + u_0 \qquad (6.437)$$

with

$$u_2 = -2d_1 n_1, \qquad u_1 = -(d_0 n_1 + 3 d_1 n_0), \qquad u_0 = -(2 d_0 n_0) \qquad (6.438)$$

Note that $R_4' = 0$ when $z = 0$ or $U = 0$. The former source of $R_4' = 0$ is a trivial case. The latter means that

$$z = \frac{-u_1 + V^{1/2}}{2 u_2} \qquad (6.439)$$

is the only real positive root, and R_4 has a minimum here.

In (6.439), the symbol V is defined as

$$V = u_1^2 - 4 u_0 u_2 \qquad (6.440)$$

which is greater that zero, in fact greater than u_1^2, z, due to the fact that $u_2 > 0$ and $u_0 < 0$ hold as a result of $n_1 < 0$, $d_1 > 0$, $d_0 > 0$, and $n_0 > 0$.

6.8. A CONCENTRATED COUPLE AT AN ARBITRARY POINT ON AN EXTERIOR SPAN OF A THREE-SPAN CONTINUOUS BEAM: GENERAL CASE

From Section 3.2.6 on the same subject, we know that the reaction R_1 is

$$R_1 = \frac{HK - FZ}{D} \qquad (6.441)$$

where

$$HK = M_0 C \frac{\left[2 C_3 L_2 + 3 \dfrac{L_3^2 - c^2}{J_3} \right]}{6} = -M_0 CA \qquad (6.442)$$

$$-FZ = CL_3 \frac{C_2 + C_3}{3} = MCB \qquad (6.443)$$

with

$$C = \frac{L_1 L_3 C_2}{6} \tag{6.444}$$

For convenience in future development, consider the expression

$$Y = \frac{R_1}{M_0} \tag{6.445}$$

6.8.1. Parametric Study of L_1

Y as a function of the parameter z under study is

$$Y = \frac{a}{(F + Gz)z} \tag{6.446}$$

where a, F, and G are constants defined as

$$a = \frac{C_2 L_3}{6}(-A + B), \qquad F = \left(\frac{L_3^2}{36}\right)(3C_2^2 + 4C_2 C_3),$$

$$G = \left(\frac{L_3^2}{9J_1}\right)(C_2 + C_3) \tag{6.447}$$

respectively.

Note that the dependency of R_1 on z is in exactly the same way as in the case of uniform load.

6.8.2. Parametric Study of L_2

Here we consider the function

$$Y = \frac{z(az + b)}{D} \tag{6.448}$$

where D is a quadratic function of z in the form

$$D = N(z^2 + hz + d) \tag{6.449}$$

with

$$h = \left(\frac{4}{3}\right)(C_1 + C_3)J_2, \qquad d = \left(\frac{4}{3}\right)C_1 C_3 J_2^2, \qquad N = \frac{L_1^2 L_3^2}{12J_2^2} \tag{6.450}$$

The constants a and b in (6.448) will be defined later.

The function Y now takes the form

$$Y = \left(\frac{1}{N}\right)\left[a + \frac{f}{g}\right]$$ (6.451)

where

$$f = (b - ah)z - ad, \qquad g = z^2 + hz + d$$ (6.452)

The sign of Y' is determined by that of P, where

$$P = (ah - b)z^2 + (2ad)z + bd$$ (6.453)

with

$$a = \left[\frac{L_1 L_3^2}{18 J_2}\right]\left(\frac{1}{J_2} - \frac{1}{J_3}\right)$$ (6.454)

$$b = -\left[\frac{L_1 L_3}{36 J_2 J_3}\right](L_3^2 - 3c^2)$$ (6.455)

It is interesting to note that both a and b can be positive, negative, or zero. For entity a, it is less than, greater than, or equal to zero when $J_3 < J_2$, $J_3 > J_2$, or $J_3 = J_2$ respectively. Thus, the sign of a is determined by the relative magnitude of J_2 and J_3, while the sign of b agrees with that of $-(L_3^2 - 3c^2)$. The former has something important to do with the relative magnitude of J values of span numbers 2 and 3 directly, and the latter ties closely to the relationship between the loaded span length and the location of the applied couple. Because of this and the resulting sign of the leading coefficient, as well as whether this coefficient is zero, it is also very important to consider the following categories:

A. $ah > b$
B. $ah = b$
C. $ah < b$

Let us look at category A first.

Category A. ah > b

Under this category, we still have several cases to consider.

Case 1. Both a and b are positive. For this case, which means that $1/J_2 > 1/J_3$ and $3c^2 > L_3^2$ simultaneously, we have $R_1' > 0$ because all coefficients in R_1' are positive. Therefore, R_1 is an increasing function of z.

Case 2. a > 0 and b < 0. For this case, which means that $1/J_2 > 1/J_3$ but $3c^2 < L_3^2$, we have one real positive root of $R_1' = 0$ (which is the same as $P = 0$), and this root corresponds to a minimum of R_1. This root is

$$z = \frac{-ad + S^{1/2}}{ah - b} \tag{6.456}$$

where

$$S = (ad)^2 - (ah - b)bd \tag{6.456A}$$

Case 3. a > 0 and b = 0. P is simplified to

$$P = ahz^2 + 2adz \tag{6.457}$$

The roots of Equation (6.457) are $z = 0$ and $z = -2d/h < 0$, both of which are not admissible. In fact, we have $P > 0$ for any nonzero z, which means that R_1 is an increasing function of z. Also, it is very interesting to observe that the entity a is the common factor in all terms in the numerator of R_1 in the present case.

Case 4. a = 0 and b < 0. We have P simplified to a quadratic equation in the special form

$$P = -bz^2 + bd \tag{6.458}$$

Now $P = 0$ when

$$z = d^{1/2} \tag{6.459}$$

and this corresponds to a minimum of R_1. Note that here the z given in (6.459) depends on d only and is independent of b and hence c also.

It is very interesting to look at (6.459) from another perspective. Since $d = 4C_1C_3J_2^2/3$, we see that z given in (6.459) can be rewritten as

$$z = AB \tag{6.459A}$$

where

$$A = \frac{\left(\dfrac{2}{3^{1/2}}\right)J_2}{(J_1 J_3)^{1/2}}, \qquad B = (L_1 L_3)^{1/2} \tag{6.459B}$$

with A as a dimensionless number, the ratio between J_2 and $(J_1 J_3)^{1/2}$ multiplied by a numerical factor, and B as the geometric mean of span lengths L_1 and L_3.

Note that $(J_1 J_3)^{1/2}$ is the geometric mean of J_1 and J_3. We also could look at (6.459) from yet another point of view and obtain

$$z = 2J_2 \left[\frac{C_1 C_3}{3} \right]^{1/2} \tag{6.459C}$$

where $(C_1 C_3)^{1/2}$ is the geometric mean of C_1 and C_3.

Case 5. a < 0 and b < 0. For this very special case, we still have a full-fledged quadratic equation $P = 0$ available. There is only one real positive root

$$z = \frac{-ad + S^{1/2}}{ah - b} \tag{6.460}$$

where

$$S = (ad)^2 - (ah - b)bd \tag{6.461}$$

Category B. ah = b

We take care of category B as follows. First of all, the leading coefficient in P is zero, so we have a linear expression for P now. Next, a and b must have the same sign or both must be zero because $h > 0$.

The special case where $a = b = 0$ means that $J_2 = J_3$ and $3c^2 = L_3^2$. This describes a specific beam with a specific relationship between the length of the loaded span and the location of the applied load, with the surprising result that $R_1 = 0$ and $P = 0$, as well as $R_1' = 0$, regardless of the magnitude of the applied concentrated couple, the parameter under study z, geometric configuration, and material properties. Of course, this is not the only way to make R_1 zero, as can be seen from assertions to be made below.

From the expression for R_1, we see that $R_1 = 0$ when

$$az + b = 0 \tag{6.462}$$

This means either a and b are zero at the same time or

$$z = \frac{-b}{a} \tag{6.463}$$

Formula (6.463) is valid as long as a and b are of opposite signs and a is nonzero.

It is plain that R_1 is positive or negative when $az + b > 0$ or < 0 respectively, and this is quite easy to realize by playing around with the values of a and b in any systematic manner. That is why only brief comments were made while doing the study on P.

Additionally, note the following observations for category B:

1. When both a and b are positive, we have $P > 0$ because a, b, and d are all positive, and as a result, all the terms in P are positive. Consequently, R_1 is an increasing function of z.
2. When both a and b are negative, we have $P < 0$, and therefore, R_1 is a decreasing function of z.

Category C. ah < b

We now treat category C where $ah < b$ in the following cases.

Case 1. a > 0 and b > 0. The equation $P = 0$ has only one real positive root given by (6.460) and (6.461) formally, with the actual a and b given under the conditions satisfied by category C.

Case 2. a < 0 and b < 0. The immediate consequence is that $P < 0$. Hence, R_1 is a decreasing functions of z. Also, we see from the expression for R_1 that R_1 is negative when both a and b are negative because D is positive.

Case 3. a < 0 and b = 0. R_1 is negative and has only one term in the numerator in the expression for R_1. Regarding P, it takes the form

$$P = ahz^2 + (2ad)z \qquad (6.464)$$

Thus, it appears that $P = 0$ when $Z = 0$ or

$$z = \frac{-2d}{h} \qquad (6.465)$$

However, neither of these roots is admissible. Therefore, (6.464) in this case implies $P < 0$. Thus, R_1 is a decreasing function of z.

Case 4. a < 0 and b > 0. Again, P is of critical importance. Here, $P = 0$ when

$$z = \frac{-ad - S^{1/2}}{ah - b} \qquad (6.466)$$

where S is as given by (6.461). The value of z given by (6.466) corresponds to a maximum R_1.

Case 5. a = 0 and b > 0. Once more, P plays the central role. We have

$$P = 0 \text{ when } z = d^{1/2} \qquad (6.467)$$

and this corresponds to a maximum R_1. Note that the z given by (6.467) depends on d only.

The cases discussed under the three main categories A, B, and C are the only admissible ones.

6.8.3. Parametric Study of L_3

We denote the parameter under study by z and consider R_1/M_0 as the subject. For convenience, let

$$\frac{R_1}{M_0} = Y \tag{6.468}$$

where

$$Y = \frac{az + bz^2 + kz^3}{D} \tag{6.469}$$

with

$$a = \frac{L_1 C_2 c^2}{12 J_3} \tag{6.470}$$

being nonnegative,

$$b = \frac{C_1 L_1 \left(C_2 - \dfrac{L_2}{J_3} \right)}{18} \tag{6.471}$$

$$k = \frac{-L_1 C_2}{36 J_3} < 0 \tag{6.472}$$

and

$$D = (d + ez)z \tag{6.473}$$

We can rewrite Y as

$$Y = \frac{k}{e} + \frac{fz + g}{z^2 + hz} \tag{6.474}$$

where

$$f = \frac{b}{e} - \frac{kd}{e^2}, \qquad g = \frac{a}{e}, \qquad h = \frac{d}{e} \tag{6.475}$$

Consider

$$G = \frac{fz + g}{z^2 + hz} = \frac{N}{H} \tag{6.476}$$

We have

$$Y = \frac{k}{e} + G \tag{6.474A}$$

$$G' = \frac{F}{H^2} \tag{6.477}$$

where

$$F = -fz^2 - 2gz - gh \tag{6.478}$$

with g nonnegative (due to the entity a being nonnegative, e being positive, and $g = a/e$), $h = d/e$ positive (due to $d > 0$ and $e > 0$), and thus $gh > 0$ or $gh = 0$.

Thus, in order to assess F, we will look into many cases regarding the sign of f and its consequences. However, it is both interesting and useful to examine the conditions under which the key player f is positive, negative, or zero and the implications before getting too busy with the interactions between f and other players.

With this in mind, we will look at $f > 0$ first. This happens if

$$\frac{b}{e} - \frac{kd}{e^2} > 0 \tag{6.479}$$

Note that even though b and k have a positive factor in common in addition to the fact that $k < 0$, it is plain that b may be positive, negative, or zero. In fact, the coefficients in the numerator of Y have the factor $L_1 C_2/36$ in common. Now when b is nonnegative, we have $-kd/(e)^2 > -b/e$, which is equivalent to (6.479). What happens if $b < 0$? Well, (6.479) is still valid. Therefore, by putting these two things together, we see that (6.479) holds for all values of b.

Let us see what happens if $f = 0$. This means that

$$eb = kd \tag{6.479A}$$

We know that both e and d are positive nonzero entities and k is strictly negative. The last equality forces us to accept the fact that $b < 0$. In other words, b can be neither zero nor positive. Now let us recall the definition of b. We see that $b < 0$ means that $C_2 < L_2/J_3$, which can be simplified to $1/J_2 < 1/J_3$. This looks very familiar. We definitely encountered this condition in an earlier section. The implications here are quite clear, and it will be left to the reader to fill in the gap.

We will finish up the examination of f by looking into the situation where $f < 0$. This inequality can be translated to mean

$$\frac{b}{e} - \frac{kd}{e^2} < 0 \tag{6.479B}$$

We call $-kd/(e)^2$ by the name K, which is positive. Then, we have $b/e + K < 0$. Thus, $b/e < -K < 0$. Consequently, $b/e < 0$, and we reached the result $b < 0$ because of $e > 0$.

Let us summarize the results from the last four paragraphs as follows. When $f > 0$, we may have all values of b, whereas either $f = 0$ or $f < 0$ is possible only when b is negative, which means that

$$\frac{1}{J_2} - \frac{1}{J_3} < 0 \tag{6.480}$$

Note that the condition $b < 0$ is just a necessary condition. Moreover, this last inequality defines a class of beams, and this class consists of two different types of beams determined by the criterion $f = 0$ or $f < 0$. Formula (6.480) is thus very powerful and has a huge impact on the results. Let us be specific about this.

For $f = 0$, the expression for F is simplified to

$$F = -2gz - gh \tag{6.481}$$

where g defined earlier is nonnegative and happens to be a common factor for the right side of (6.481). Thus, $F = 0$, < 0, or > 0 when $-(2z + h) = 0$, < 0, or > 0 respectively. Therefore, $F = 0$ when $z = -h/2$, and this is meaningful only when $h < 0$. However, we know that, by definition, $h > 0$, so it is impossible to have $F = 0$. Moreover, because z, g, and h are all positive, we see from the appearance of F exhibited in (6.481) that $F < 0$. This means that G is a decreasing function of z and so is Y, and eventually R_1 is a decreasing function of z.

Let us return to $f = 0$ itself to see the meaning of this situation. It means, first of all, that $b = kd/e$. That is,

$$\frac{J_2}{J_3} = \frac{A}{B} \tag{6.482}$$

where

$$A = 12C_1 + 11C_2, \qquad B = 8(C_1 + C_2) \tag{6.483}$$

with A and B both positive and simple linear algebraic functions of C_1 and C_2 only.

Utilizing these notations introduced in (6.483), we may rewrite the statements for $f < 0$ and $f > 0$ separately as follows. For the case where $f < 0$, we have the defining property

$$\frac{J_2}{J_3} > \frac{A}{B} \tag{6.484}$$

For the case where $f > 0$, we have

$$\frac{J_2}{J_3} < \frac{A}{B} \tag{6.485}$$

Thus, from (6.482), (6.484), and (6.485), J_2/J_3 becomes a convenient index for handy reference and usage.

Now let us examine all the possible cases regarding the coefficients in F and their effects on F and Y.

Case 1. f > 0 and g > 0

As a direct result of $g > 0$ and the given definition of h, we have $gh > 0$. This means that all the coefficients in F are negative, in addition to z itself being positive. Thus, $F < 0$, $G' < 0$, and $Y' < 0$, and as a result Y is a decreasing function of z. But what do we mean by $g > 0$? Since $g = a/b$ by definition and $e > 0$, we know that $g > 0$ means $a > 0$. From the definition of a, the inequality $a > 0$ means that c is nonzero and positive of course.

Note that the expression for Y consists of two terms. One is k/e, which is independent of z, and the other is G, which is a function of z. We see that $k/e < 0$ because of $k < 0$. The function G, on the other hand, can be positive and is in fact positive for the present case. Thus, it is conceivable that Y may become zero, negative, or positive. It would be interesting to explore in more detail along this line.

Case 2. f > 0 and g = 0

From $g = 0$, we have $gh = 0$ and F becomes $F = -fz^2$, which is less than zero, signifying a negative first derivative of G. Thus, G is a decreasing function of z and so is Y. Note that the function G itself is simplified to

$$G = \frac{fz}{z^2 + hz} \tag{6.486}$$

which is positive because both f and h are positive, in addition to z itself being positive.

Once again, for the same reason stated in the last paragraph of case 1, Y can be positive, negative, or zero depending on the relative magnitude of G and k/e.

Case 3. f = 0 and g > 0

The immediate consequence of $g > 0$ is $gh > 0$. Thus, with $f = 0$, we have $F = -2gz - gh < 0$ and G is a decreasing function of z. The fact that $f = 0$ gives us a simplified G as

$$G = \frac{g}{z^2 + hz} \tag{6.487}$$

which is positive.

The same remarks regarding the sign of Y that appeared in the last paragraph of case 1 apply here too. Also note the requirement for $f = 0$ in terms of J_2, J_3 and A, B as expressed in (6.482).

Case 4. $f = 0$ and $g = 0$

We have immediately $F = 0$ and $G = 0$ also. Thus, $Y = k/e < 0$ and Y is independent of z.

Note here that $a = 0$ and $b < 0$ as a result of $g = 0$ and $f = 0$ respectively. Thus, this is a class of beams with the concentrated couple applied at zero distance from the interior support of the exterior span with the property stipulated by $b < 0$.

The same remark embedded in the last sentence of case 3 applies here as well.

Case 5. $f < 0$ and $g = 0$

Because of $g = 0$, we have $F = -fz^2$, which is positive due to the given fact that $f < 0$. Thus, G is an increasing function of z. Here again, we have $a = 0$ and $b < 0$. The function G is reduced to

$$G = \frac{f}{z + h} \tag{6.488}$$

which is negative, because f is negative, and h and z are positive. The function Y is therefore always negative and its absolute value increases as z decreases.

Case 6. $f < 0$ and $g > 0$

We have $gh > 0$ and

$$F = -fz^2 - 2gz - gh \tag{6.489}$$

which can be positive, negative, or zero.

For $F = 0$, we have the only real positive root

$$z = \frac{g + S^{1/2}}{-f} \tag{6.490}$$

where

$$S = g^2 - fgh \tag{6.491}$$

G has a minimum here. Note that again $b < 0$ here is required by $f < 0$.

6.8.4. Parametric Study of J_1

In order to change pace, let us denote the parameter by $1/x$. Thus, for the purpose of studying R_1, let us consider a function Y of x defined by (6.492) below. In other words, in order to study the effects of J_1, henceforth denoted by $1/x$, on reaction R_1, it is sufficient to scrutinize R_1/M_0, represented by the function

$$Y = \frac{C}{A + Bx} \tag{6.492}$$

where A, B, and C are constants defined by

$$A = \left[\frac{L_1 L_3}{6}\right]^2 [3C_2^2 + 4C_2 C_3] \tag{6.493}$$

$$B = 4\left[\frac{L_1 L_3}{6}\right]^2 L_1[C_2 + C_3] \tag{6.494}$$

$$C = \left(\frac{L_1 L_3 C_2}{6}\right)(S - T) \tag{6.495}$$

with

$$S = \frac{L_3(C_2 + C_3)}{3} \tag{6.496}$$

$$T = \frac{2C_3 L_2 + 3(L_3^2 - c^2)}{6} \tag{6.497}$$

Note that A, B, S, and T are all positive, whereas C may be positive, negative, or zero depending on the relative magnitude of S and T, since the other factor of C is positive.

Now

$$Y' = -CB(A + Bx)^{-2} \tag{6.498}$$

is positive, negative, or zero when C is negative, positive, or zero respectively due to the fact that all the other factors of Y' are positive, but capped with a minus sign.

From (6.495), we see that $C = 0$ when $S = T$, whereas $C < 0$ when $S < T$, and $C > 0$ when $S > T$.

From (6.498), we see that $Y' = 0$ when $C = 0$. However,

$$Y'' = 2CB(A + Bx)^{-3} \qquad (6.499)$$

so $Y'' = 0$ at $C = 0$. Hence $C = 0$ is a point of inflection.

When $C < 0$, we have $Y' > 0$ and Y as an increasing function of x. When $C > 0$, we have $Y' < 0$ and Y as a decreasing function of x.

Note that S is a function of L_3, C_2, and C_3, while T is a function of L_2, L_3, C_3, and c. Therefore, the conditions $C = 0$, $C < 0$, and $C > 0$ represent relationships among C_2, C_3, L_2, L_3, and c. We see that C is the key to the behavior of Y represented by various features of C being zero, positive, or negative and an increasing or decreasing function of z or stationary. Moreover, we can pick up one particular entity and state it in terms of the other entities in an expression resulting from $C = 0$, $C < 0$, or $C > 0$. By doing so, there exists a huge space for us to explore.

6.8.5. Parametric Study of J_2

We will denote $1/J_2$ by x. Then the function representing R_1/M_0 is Y defined as

$$Y = \frac{ax^2 + bx}{D} \qquad (6.500)$$

where a and b are constants and D is defined as

$$D = kx^2 + (xd) + e \qquad (6.501)$$

with k, d, and e as constants.

We see that (6.500) is basically the same mathematical problem as the one solved in the study of L_2. The only thing that is different is the set of expressions for the constants a, b, k, d, and e, which are as follows:

$$a = -\left(\frac{L_2^2 L_3}{3}\right)A \qquad (6.502)$$

$$b = A\left(\frac{-L_2 L_3 C_3}{3} + L_2 B\right) \qquad (6.503)$$

$$A = \frac{L_1 L_3}{6}, \qquad B = \frac{2L_2 L_3 + 3(L_3^2 - c^2)}{6 J_3} \qquad (6.504)$$

$$k = 3A^2 L_2^2 \qquad (6.505)$$

$$d = 4A^2(C_1 + C_3)L_2 \qquad (6.506)$$

$$e = 4A^2 C_1 C_3 \qquad (6.507)$$

6.8.6. Parametric Study of J_3

We denote the parameter J_3 to be studied by $1/x$ for convenience. The function representing R_1/M_0 is Y, which is defined as

$$Y = \frac{ax + b}{D} \tag{6.508}$$

where

$$a = A\left(\frac{L_3^2}{3} - B\right) \tag{6.509}$$

$$b = \frac{AL_3C_2}{3} > 0 \tag{6.510}$$

$$D = kx + d \tag{6.510A}$$

with

$$k = 4\left(\frac{L_1L_3}{6}\right)^2 SL_3 > 0 \tag{6.511}$$

$$d = \left[\frac{L_1L_3}{6}\right]^2 T > 0 \tag{6.512}$$

and

$$A = \frac{L_1L_3C_2}{6} \tag{6.513}$$

$$B = \frac{2L_2L_3 + 3(L_3^2 - c^2)}{6} \tag{6.514}$$

$$S = C_1 + C_2 \tag{6.515}$$

$$T = 3C_2^2 + 4C_1C_2 \tag{6.516}$$

For convenience in future use, we also introduce

$$T_1 = \frac{T}{C_2} \tag{6.517}$$

The function Y can be rewritten as

$$Y = \frac{a}{k} + F \tag{6.518}$$

where

$$F = \frac{f}{x + e} \tag{6.519}$$

$$e = \frac{d}{k}, \qquad f = \frac{bk - ad}{k^2} \tag{6.520}$$

Then, we have

$$Y' = \frac{-f}{(x + e)^2} \tag{6.521}$$

and $Y' = 0, < 0,$ or > 0 when $f = 0, > 0,$ or < 0 respectively because the denominator in (6.521) is positive, capped with a minus sign. That is,

$$Y' = 0 \text{ when } bk - ad = 0 \tag{6.522}$$

$$Y' < 0 \text{ when } bk - ad > 0 \tag{6.523}$$

$$Y' > 0 \text{ when } bk - ad < 0 \tag{6.524}$$

Now $bk - ad$ can be rewritten as

$$bk - ad = CC_2\left[\frac{4L_3^2 C_2 S}{3} - \left(\frac{L_3^2}{3} - B\right)T\right] \tag{6.525}$$

with

$$C = \left(\frac{L_1 L_3}{6}\right)^3 \tag{6.525A}$$

It can be seen clearly from (6.525) that

$$f = 0 \text{ when } 4C_2 L_3^2 S = (L_3^2 - 3B)T \tag{6.526}$$

$$f > 0 \text{ when } 4C_2 L_3^2 S > (L_3^2 - 3B)T \tag{6.527}$$

$$f < 0 \text{ when } 4C_2 L_3^2 S < (L_3^2 - 3B)T \tag{6.528}$$

It is interesting to look at the cases where $a = 0, a > 0,$ and $a < 0.$ For $a = 0,$ $bk - ad = bk = (4/3)CC_2^2 L_3^2 S > 0.$ This means that $f > 0, Y' > 0,$ and Y is a decreasing function of $x.$ For $a > 0,$ we have $L_3^2/3 - B > 0.$ This means that $3c^2 > 2L_2 L_3 + L_3^2.$ Since $L_3 > c,$ we have here $3L_3^2 > 3c^2 > 2L_2 L_3 + L_3^2.$ Thus, $L_3 > L_2.$ For $a < 0,$ we have from $L_3^2/3 - B < 0$ the result $3c^2 < 2L_2 L_3 + L_3^2.$

Let us examine (6.526) in greater detail by opening up the package labeled $B.$ Thus, $L_3 > L_2.$ The result is

$$L_3^2\left(\frac{4S}{3} + \frac{T_1}{6}\right) + \left(\frac{T_1 L_2}{3}\right)L_3 = \frac{T_1 c^2}{2} \tag{6.529}$$

which can be solved for L_3 in terms of the other entities if we so desire. For convenience in future use, let us call the left side of (6.529) by the name L. Thus, (6.529) is

$$L = \frac{T_1 c^2}{2} \qquad (6.530)$$

For the case where $f > 0$, we have

$$L > \frac{T_1 c^2}{2} \qquad (6.531)$$

and the necessary condition for $f < 0$ to hold is naturally

$$L < \frac{T_1 c^2}{2} \qquad (6.532)$$

Now let us review the expression L and see what entities are affecting the value of L. We see that S, T, L_2, L_3, and C_2 are involved. By definition, S and T are from D directly and L is an increasing function of these two entities.

Formulae (6.530–6.532) can be rewritten respectively as

$$4SL_3^2 = \frac{3T_1 a}{A} \qquad (6.533)$$

$$4SL_3^2 > \frac{3T_1 a}{A} \qquad (6.534)$$

$$4SL_3^2 < \frac{3T_1 a}{A} \qquad (6.535)$$

We can see clearly from these last three formulae the effects of a/A on the entity a itself. For (6.533), the entity a can only be positive because all the other factors in this equality are positive. For (6.535), the left-hand side is positive and is less than the right-hand side; therefore, the right-hand side cannot be zero or negative. In other words, the right-hand side must be positive as a start. As for (6.534), the situation is much more relaxed in that the entity a can be any real number, positive, negative, or zero.

When we look at the last three formulae from another perspective, we see something very interesting happening. If we divide both sides of these formulae by the positive entity T_1, we have on the left-hand side $4(S/T_1)L_3^2$, whereas the right-hand side becomes simply $3a/A$. Thus, these formulae provide us with a set of criteria for comparison purposes. Of course, we could divide both sides of these formulae by the positive entity S instead of what we just did.

Another interesting point is the effects of the entity a on f. Recall that the entity a can be positive, negative, or zero and from the way the entity a appears in f, we see that f is a decreasing function of the entity a. Let us look at $f = 0$ for possible minimum or maximum values of the function Y. We note that the second term of Y, called F, which contains x, vanishes when $f = 0$. In fact, both F' and F'' vanish when $f = 0$, so we have a point of inflection when $f = 0$. It is clear from the definition of Y that Y can be positive, negative, or zero depending on the relative magnitude of a/k and F as well as the signs of the entities a and f.

6.9. UNIFORM LOAD ON THE INTERIOR SPAN OF A THREE-SPAN CONTINUOUS BEAM: GENERAL CASE

The subject under study is the reaction R_1 given by

$$R_1 = wSL_3Z_1 \tag{6.536}$$

where

$$S = \frac{L_1L_3}{36J_2L_2D} \tag{6.537}$$

$$D = \left[\frac{L_1L_3}{6}\right]^2 [3C_2^2 + 4(C_1 + C_3)C_2 + 4C_1C_3] \tag{6.538}$$

$$Z_1 = (A_0L_2 + A_1L_2^2)E_2 - (2A_1L_2 + A_0)E_3 + A_1E_4 \tag{6.539}$$

with

$$A_0 = -2L_2C_3 \tag{6.540}$$

$$A_1 = -(3C_2 + 2C_3) \tag{6.541}$$

$$E_j = \frac{e^j - a^j}{j}, \quad j = 2, 3, 4 \tag{6.542}$$

For convenience, S may be rewritten as

$$S = \frac{1}{J_2L_1L_2L_3K} \tag{6.543}$$

where

$$K = [3C_2^2 + 4(C_1 + C_3)C_2 + 4C_1C_3] \tag{6.544}$$

Substituting (6.540–6.542) into (6.539), we have

$$Z_1 = -(3C_2 + 4C_3)L_2^2E_2 + 6(C_2 + C_3)L_2E_3 + A_1E_4 \tag{6.545}$$

Substituting (6.543) into (6.536), we obtain

$$\frac{R_1}{w} = \frac{Z_1}{J_2L_1L_2K} \tag{6.546}$$

For $w = $ a constant, it is expedient to consider R_1/w instead of R_1 itself. We see that R_1/w is basically a function of J_i, L_i $(i = 1, 2, 3)$ when e and a are constants. We also can consider R_1/w as a function of J_i, L_i $(i = 1, 2, 3)$ as well as e and a.

6.9.1. Parametric Study of L_1

Denoting the parameter by z, we have the following. Z_1 is a constant.

$$L_1 K = z(a_0 + a_1 z) \tag{6.547}$$

where

$$a_0 = 3C_2^2 + 4C_2 C_3 > 0 \tag{6.548}$$

$$a_1 = 4 \left(\frac{C_2 + C_3}{J_1} \right) > 0 \tag{6.549}$$

For convenience, let us denote R_1/w by Y. Thus,

$$Y = \frac{A}{z(a_0 + a_1 z)} \tag{6.550}$$

where

$$A = \frac{Z_1}{J_2 L_2} \tag{6.551}$$

Observing (6.550), we see that Y is a decreasing function of z when A is positive. That is,

$$Y' = -A[z(a_0 + a_1 z)]^{-2} B < 0 \text{ with } B = a_0 + 2a_1 z > 0 \tag{6.552}$$

when A is positive. But what about the cases where A is zero or negative?

When $A = 0$, $Y' = 0$. What about Y'' when $A = 0$? Well, $A = 0$ also means that $Y'' = 0$, so $A = 0$ is a point of inflection. Naturally, when $A < 0$, we have $Y' > 0$, meaning that Y is an increasing function of z. As to the detailed descriptions of the conditions $A < 0$, $A = 0$, and $A > 0$, we will see them below.

Moreover, what are a_0 and a_1 doing here? We see that the larger the values of a_0 and a_1 are, the smaller the value of Y for a given z becomes. Also, from (6.548) and (6.549), a_0 and a_1 increase with increasing C_2 and C_3, in addition to the fact that a_1 is inversely proportional to J_1.

Now, what about A? From (6.551), A is proportional to Z_1 and inversely proportional to J_2 and L_2. Thus, for a given z, the larger the values of J_2 and L_3 are, the smaller the value of A, and thus that of Y, becomes.

What about Z_1? We will pursue it as follows.

$$Z_1 = -(L_2^2 E_2 m_2 + E_4 m_4) + m_3 E_3 L_2 \qquad (6.553)$$

where

$$m_2 = (3C_2 + 4C_3) > 0 \qquad (6.554)$$

$$m_3 = 6(C_2 + C_3) > 0 \qquad (6.555)$$

$$m_4 = (3C_2 + 2C_3) > 0 \qquad (6.556)$$

By (6.553), we have

$$Z_1 < 0, \; Z_1 = 0, \; \text{or} \; Z_1 > 0 \qquad (6.557)$$

when

$$m_3 E_3 L_2 < E_2 m_2 L_2^2 + E_4 m_4 \qquad (6.558)$$

or

$$m_3 E_3 L_2 = E_2 m_2 L_2^2 + E_4 m_4 \qquad (6.559)$$

or

$$m_3 E_3 L_2 > E_2 m_2 L_2^2 + E_4 m_4 \qquad (6.560)$$

respectively.

Note that depending on the values of L_2, L_3, J_2, J_3, and E_j ($j = 2, 3, 4$), one of these last three formulae holds for a given situation. Since Z_1 and Y are so related that one is proportional to the other, we see that $Y < 0$, $Y = 0$, or $Y > 0$ if and only if $Z_1 < 0$, $Z_1 = 0$, or $Z_1 > 0$ respectively. Hence, we can call Z_1 the "character indicator" for Y.

6.9.2. Parametric Study of L_2

As usual, we will denote the parameter by z. Again, Z_1 is the character indicator. It is expedient to consider (6.546). We will deal with K first. Thus,

$$K = k_0 + k_1 z + k_2 z^2 \qquad (6.561)$$

where

$$k_0 = 4C_1 C_3 > 0, \qquad k_1 = \frac{4(C_1 + C_3)}{J_2} > 0, \qquad k_2 = \frac{3}{J_2^2} > 0 \quad (6.562)$$

Then we look at Z_1 and have

$$Z_1 = b_3 z^3 + b_2 z^2 + b_1 z + b_0 \qquad (6.563)$$

where

$$b_3 = \frac{-3E_2}{J_2}, \qquad\qquad b_2 = -4C_3E_2 + \frac{6E_3}{J_2},$$

$$\tag{6.564}$$

$$b_1 = 6C_3E_3 - \frac{3E_4}{J_2}, \qquad b_0 = -2C_3E_4 < 0$$

Thus,

$$\frac{R_1}{w} = Y = \frac{Z_1}{J_2 L_1 M} \tag{6.565}$$

where

$$M = z(k_2 z^2 + k_1 z + k_0) \tag{6.566}$$

To facilitate the subsequent analysis, we will rewrite Y as

$$Y = \left[\frac{1}{J_2 L_1 k_2}\right]\left[b_3 + \frac{N}{z^3 + a_2 z^2 + a_1 z}\right] \tag{6.567}$$

where

$$N = n_2 z^2 + n_1 z + b_0 \tag{6.568}$$

with

$$n_2 = b_2 - a_2 b_3, \qquad n_1 = b_1 - a_1 b_3, \qquad a_2 = \frac{k_1}{k_2}, \qquad a_1 = \frac{k_0}{k_2} \tag{6.569}$$

It is very interesting to note that $b_2 < 0$, $= 0$, or > 0 if and only if

$$\frac{3E_3}{J_2} <, =, \text{ or } > 2C_3E_2 \tag{6.570}$$

respectively. Also, $b_1 < 0$, $= 0$, or > 0 when

$$2C_3E_3 <, =, \text{ or } > \frac{E_4}{J_2} \tag{6.571}$$

respectively.

In both (6.570) and (6.571), only C_3, J_2, and E_2, E_3, E_4 are involved. From (6.569), we see that

$$n_2 < 0, = 0, \text{ or } > 0 \text{ when } b_2 <, =, \text{ or } > a_2 b_3 \tag{6.572}$$

respectively and

$$n_1 < 0, = 0, \text{ or } > 0 \text{ when } b_1 <, =, \text{ or } > a_1 b_3 \tag{6.573}$$

respectively. Also, note that $a_j > 0$ for $j = 1, 2$, but $b_3 < 0$. Therefore,

$$-a_2 b_3 > 0, \qquad -a_1 b_3 > 0 \tag{6.574}$$

Expressions in (6.574) are important for evaluating n_2 and n_1 needed in N.

Consider the case where $N = 0$. This means that

$$z = \frac{-n_1 + S^{\frac{1}{2}}}{2n_2} \tag{6.575}$$

or

$$z = \frac{-n_1 - S^{\frac{1}{2}}}{2n_2} \tag{6.576}$$

where

$$S = n_1^2 - 4n_2 b_0 \tag{6.577}$$

Note that $-4n_2 b_0 > 0$ in (6.577) if $n_2 > 0$ because it is given that $b_0 < 0$, so $S > 0$ and in fact $S^{\frac{1}{2}}$ is greater than the absolute value of n_1. Therefore, there are two real roots, but there may be only one that is admissible, and we have to investigate case by case regarding the sign of n_2 and n_1. The details about this kind of investigation can be found in Section 6.3 and will not be repeated here. Rather, we will see the conditions and implications of the sign of the coefficients of N.

Case 1. $n_2 = 0$

This happens when $b_2 = a_2 b_3$. This condition is equivalent to

$$-4C_3 E_2 + \frac{6E_3}{J_2} = \left(\frac{k_1}{k_2}\right)\left[\frac{-3E_2}{J_2}\right] \tag{6.578}$$

$$E_2\left[\frac{3k_1}{k_2 J_2} - 4C_3\right] + \frac{6E_3}{J_2} = 0 \tag{6.578A}$$

It appears that this is perfectly harmless, and it is therefore quite possible to pick up suitable C_2, C_3, J_2, E_2, and E_3 to satisfy the equality specified in (6.578A). Furthermore, this is supposed to be valid in general in that N can be positive, negative, or zero. (Well, is it really so?) However, when $N = 0$, we have right away $N = n_1 z + b_0 = 0$ in the present case, which leads to

$$z = \frac{2C_3 E_4}{n_1} \tag{6.579}$$

where

$$n_1 = 4C_1 C_3 J_2 E_2 + 6C_3 E_3 - \frac{3E_4}{J_2} \qquad (6.580)$$

This is beautiful, and it appears that it can take all real values and is, therefore, fine.
Next, let us look directly at

$$n_2 = b_2 - a_2 b_3 = 4C_1 E_2 + \frac{6E_3}{J_2} \qquad (6.581)$$

which is clearly positive in general except when $E_2 = E_3 = 0$, meaning $a = e$, which
is a trivial case.

The moral of the story is that we should have dug a little deeper before we
stopped at (6.578A); then we would have reached (6.581). The above is written
primarily with the beginner in mind; however, even the seasoned reader may benefit
from this in at least two aspects: recreation and precaution.

Since we are studying $n_2 = 0$, we might just as well do $n_1 = 0$ also. We look at
(6.580) and see that this happens when

$$4C_1 C_3 J_2 E_2 + 6C_3 E_3 = \frac{3E_4}{J_2} \qquad (6.582)$$

Let us continue with n_1. We note that $n_1 > 0$ when

$$4C_1 C_3 J_2 E_2 + 6C_3 E_3 > \frac{3E_4}{J_2} \qquad (6.583)$$

and $n_1 < 0$ when

$$4C_1 C_3 J_2 E_2 + 6C_3 E_3 < \frac{3E_4}{J_2} \qquad (6.584)$$

Now we return to consideration of $N < 0$ and $N > 0$. To this end, let us denote
the two possible roots of $N = 0$ by z_1 and z_2 with $z_1 < z_2$. Then, since $n_2 > 0$, we
have $N > 0$ if and only if

$$(z - z_1)(z - z_2) > 0 \qquad (6.585)$$

meaning $z > z_1$ and $z > z_2$ simultaneously or $z < z_1$ and $z < z_2$ simultaneously. The
former means that $z > z_2$ and the latter signifies that $z < z_1$.

Similarly, $N < 0$ if and only if

$$(z - z_1)(z - z_2) < 0 \qquad (6.586)$$

meaning $z < z_2$ and $z > z_1$ simultaneously. In other words, we have $z_1 < z < z_2$.

The consequence of $n_1 = 0$ is a much more simplified N with its implications:

$$N = n_2 z^2 + b_0 \qquad (6.587)$$

If $N = 0$ additionally, then we have

$$z = \left[\frac{-b_0}{n_2}\right]^{1/2} = \left[\frac{C_3 E_4}{2C_1 E_2 + \dfrac{3E_3}{J_2}}\right]^{1/2} \tag{6.588}$$

If $N > 0$ instead, then we have

$$z > \left(\frac{-b_0}{n_2}\right)^{1/2} \tag{6.589}$$

If $N < 0$ is the case, then we have

$$z < \left(\frac{-b_0}{n_2}\right)^{1/2} \tag{6.590}$$

When we look at (6.567), we notice that Y is the product of two factors, $1/(J_2 k_2 L_1)$ and $(b_3 + N/F)$, where

$$F = z^3 + a_2 z^2 + a_1 z \tag{6.591}$$

is a third-degree polynomial in a positive variable z with all positive coefficients and is therefore always positive. N, on the other hand, as we notice from above, may be positive, negative, or zero, whereas b_3 is negative. Hence, negative N will enhance the negative value of Y along with the intrinsic negative value of b_3.

In order to change pace, let us look at some interesting results. If $b_2 = 0$, then

$$\frac{3E_3}{J_2} = 2E_2 C_3 \tag{6.592}$$

and

$$n_2 = -a_2 b_3 > 0 \tag{6.592A}$$

If $b_1 = 0$, then

$$2C_3 E_3 = \frac{E_4}{J_2} \tag{6.593}$$

and

$$n_1 = -a_1 b_3 > 0 \tag{6.593A}$$

If both b_1 and b_2 are zero, then both (6.592) and (6.593) hold, and as a result, we have

$$3E_4 = (2C_2 J_3)^2 E_2 \tag{6.594}$$

For the sake of changing pace and gathering more information, let us explore the following approach. Consider the equality

$$\frac{N}{z(z^2 + a_2 z + a_1)} = \frac{A}{z} + \frac{Bz + C}{z^2 + a_2 z + a_1} \tag{6.595}$$

Thus,

$$n_2 z^2 + n_1 z + b_0 = A(z^2 + a_2 z + a_1) + (Bz + C)z \tag{6.596}$$

We have

$$n_2 = A + B, \qquad n_1 = Aa_2 + C, \qquad b_0 = Aa_1 \tag{6.597}$$

and

$$A = \frac{b_0}{a_1}, \qquad B = n_2 - \frac{b_0}{a_1}, \qquad C = n_1 - \frac{a_2 b_0}{a_1} \tag{6.598}$$

We may do an exploration using a procedure similar to the two-span case outlined earlier.

6.9.3. Parametric Study of L_3

The function to be studied is

$$Y = \frac{Z_1}{J_2 L_1 L_2 K} \tag{6.599}$$

where

$$K = k_1 z + k_0 \tag{6.600}$$

with z as the independent variable and

$$k_1 = 4\left(\frac{C_1 + C_2}{J_3}\right) > 0, \qquad k_0 = 3C_2^2 + 4C_1 C_2 > 0 \tag{6.601}$$

Also,

$$Z_1 = b_1 z + b_0 \tag{6.602}$$

where

$$b_1 = 2\left[\frac{-2L_2^2 E_2 + 3L_2 E_3 - E_4}{J_3}\right] \tag{6.603}$$

$$b_0 = 3C_2[-L_2^2 E_2 + 2L_2 J_3 E_3 - E_4] \tag{6.604}$$

Let us focus on

$$\frac{Z_1}{K} = \left(\frac{1}{k_1}\right)\left[b_1 + \frac{N}{M}\right] \tag{6.605}$$

where

$$M = z + \frac{k_0}{k_1} \tag{6.606}$$

$$N = b_0 - \frac{b_1 k_0}{k_1} \tag{6.607}$$

Note that

$$\frac{k_0}{k_1} = \frac{(3C_2 + 4C_1)(C_2 J_3)}{4(C_1 + C_2)} > 0 \tag{6.608}$$

Therefore, $M > 0$, while N determines the sign of N/M in (6.605).

Note also that increasing z in (6.605) implies decreasing absolute value of N/M in all the cases to be presented below. Among other things, whether b_1 and b_0 can be positive, negative, or zero warrants consideration of the various cases in the following paragraphs.

Case 1. $b_1 = 0$ and $b_0 = 0$

The immediate consequence is $zn = 0$, which leads to $Z_1 = 0$ and $Y = 0$. Moreover, in terms of case-defining conditions, we have the following.

$b_1 = 0$ means

$$3L_2 E_3 = E_4 + 2L_2^2 E_2 \tag{6.609}$$

which is an equality connecting L_2 and E_j ($j = 2, 3, 4$). In other words, it is a simple quadratic equation in L_2, with numerical multiples of E_j ($j = 2, 3, 4$) as coefficients if we want to view it that way.

$b_0 = 0$ means

$$2L_2 E_3 = (L_2^2 E_2 + E_4) \tag{6.610}$$

It is interesting to note that by combining (6.609) and (6.610) we obtain the interesting and simple result

$$E_3 = L_2 E_2 \tag{6.611}$$

which signifies a very special span length L_2, namely

$$L_2 = \frac{E_3}{E_2} \tag{6.612}$$

This span length has an interesting upper bound:

$$L = \frac{2(e + a)}{3} \qquad (6.613)$$

Moreover, if we combine (6.610) and (6.611), we will obtain

$$L_2 E_3 = E_4 = L_2^2 E_2 \qquad (6.611A)$$

Now let us focus on (6.611A). We see the very interesting result

$$L_2 = \frac{E_4}{E_3} \qquad (6.612A)$$

Comparing (6.612) and (6.612A), we arrive at the simple, concise, and wonderful conclusion that

$$\frac{E_4}{E_3} = \frac{E_3}{E_2} \qquad (6.612B)$$

characterizes this case.

Case 2. $b_1 = 0$ and $b_0 < 0$

It is interesting to observe that only L_2 and E_j appear in all the formulae in the present case except (6.613), where even though L_2 and E_j do not show up explicitly, L_2 and the elements that make up E_j are specifically referred to.

The first defining condition of the present case is stated in (6.609), whereas the second one says

$$2L_2 E_3 < L_2^2 E_2 + E_4 \qquad (6.614)$$

Combining (6.609) and (6.614), we have $L_2 E_3 > L_2^2 E_2$, which is equivalent to

$$E_3 > L_2 E_2 \qquad (6.615)$$

Here again, (6.615) says that we have case 2 when the actual span length L_2 is less than the special span length defined in (6.612). Therefore, this special span length is really critical.

What are the impacts of the definition of the present case on the function Y without resorting to Y', the first derivative of Y with respect to z? First, note that the larger the absolute value of b_0 is, the larger the absolute value of N becomes. Since $b_1 = 0$, we have that Y is proportional to N, which is, in the present case, equal to b_0. Also, of course, Y is negative here because N is. By mere observation, we see that Y is an increasing function of z due to the fact that N is a negative constant and the way that the single term z appears in the denominator of the expression for Y.

A point to emphasize is that Y is proportional to b_0. Let us return to the definition of b_0. We see from (6.604) that

$$b_0 = 3C_2P \qquad (6.616)$$

where

$$P = -E_2L_2^2 + 2E_3L_2 - E_4 \qquad (6.617)$$

will be seen as something of fundamental importance, and that is the reason why we coin a special symbol for it. From (6.616), we see that C_2 and P are proportional to b_0. Let us look at P closely. Here, since $C_2 > 0$, we see that $b_0 < 0$ is due to $P < 0$.

Now consider $U = -P$ as a function of L_2. Thus

$$U = E_2L_2^2 - 2E_3L_2 + E_4 \qquad (6.618)$$

and

$$U' = 2E_2L_2 - 2E_3 \qquad (6.619)$$

$$U'' = 2E_2 > 0 \qquad (6.620)$$

$$U' = 0 \text{ when } L_2 = \frac{E_3}{E_2} \qquad (6.621)$$

Hence, L_2 defined in (6.621), which is the same as the definition in (6.612), corresponds to a minimum value of U.

We see, then, that for a given set of E_j ($j = 2, 3, 4$), U and P vary as L_2 varies and P has its maximum at the L_2 given in (6.621), where U has its minimum. Note also that in the present case, only one span length L_2 is involved in any of the formulae here.

Case 3. $b_1 = 0$ and $b_0 > 0$

Using a procedure similar to the one in case 2, we have for the second condition above, $b_0 > 0$,

$$2L_2E_3 > L_2^2E_2 + E_4 \qquad (6.622)$$

which, along with the defining condition of $b_1 = 0$ displayed in (6.609), yields the simple result

$$E_3 < L_2E_2 \qquad (6.623)$$

Note that there are big differences too; for example, $N > 0$ here and so $Y > 0$. Remarks similar to those at the end of case 1, without reference to (6.613), apply here. The rest of the remarks and conclusions are analogous to those in case 2 and will, therefore, be left for the interested reader to fill in.

Case 4. $b_1 > 0$ and $b_0 = 0$

The first defining condition means that

$$3L_2E_3 > E_4 + 2L_2^2E_2 \tag{6.624}$$

which, along with the second defining condition displayed earlier in case 1, gives us

$$E_3 > L_2E_2 \tag{6.625}$$

The expression N is simplified to

$$N = \frac{-b_1k_0}{k_1} < 0 \tag{6.626}$$

because k_0, k_0, and b_1 are all positive.

As far as the N/M part of Y is concerned, the situation here is quite similar to that in case 2. However, in view of (6.626) as well as the expression for Y and the fact that $b_1 > 0$, we see that Y may be positive, negative, or zero.

Remarks similar to those at the end of case 2 apply here.

Case 5. $b_1 > 0$ and $b_0 < 0$

The first and second defining conditions here are given in (6.624) and (6.614) respectively. For direct consequences of these defining conditions, see the appropriate paragraphs above.

Now the impacts of these conditions on N and, therefore, on Y can be seen from

$$N = b_0 - \frac{b_1k_0}{k_1} < 0 \tag{6.626A}$$

because both terms in N are negative. Note again that Y has two terms, one of which involves N and the other which does not, and that these two terms are of opposite signs. Hence, Y may be positive, negative, or zero, and the possibilities are just like the last case.

Case 6. $b_1 > 0$ and $b_0 > 0$

The first defining condition of the case leads to

$$3L_2E_3 > E_4 + 2L_2^2E_2 \tag{6.627}$$

The second defining condition yields

$$2L_2E_3 > E_4 + L_2^2E_2 \tag{6.628}$$

It is very interesting to note the resemblance between these last two inequalities.

The conclusion here is that N as well as Y may be positive, negative, or zero. Note that, once again, formulae here involve L_2 and E_j only, with $j = 2, 3, 4$.

Case 7. $b_1 < 0$ and $b_0 = 0$

This happens when

$$3L_2E_3 < E_4 + 2L_2^2E_2 \tag{6.629}$$

and

$$2L_2E_3 = L_2^2E_2 + E_4 \tag{6.630}$$

are satisfied simultaneously.

It is fascinating to note that subtracting (6.630) from (6.629) yields $L_2E_3 < L_2^2E_2$ or simply

$$E_3 < L_2E_2 \tag{6.631}$$

which is exactly as (6.623) appeared in case 3. It is incredible to have the same governing formulae in two different cases where the big difference lies in the sign of Y, as will be noted below,

Note that $Z_1 = b_1z < 0$ because $b_1 < 0$ and $b_0 = 0$, but $N = -b_1k_0/k_1 > 0$. Thus, $Y < 0$, $Y = 0$, and $Y > 0$ are all possible. Also, $Y' < 0$ is the situation here, signifying Y as a decreasing function of z.

Case 8. $b_1 < 0$ and $b_0 < 0$

The defining conditions are

$$3L_2E_3 < E_4 + 2L_2^2E_2 \tag{6.632}$$
$$2L_2E_3 < E_4 + L_2^2E_2 \tag{6.633}$$

respectively. N may be positive, negative, or zero, and this is similar to case 6. As a result, Y may be positive, negative, or zero.

Case 9. $b_1 < 0$ and $b_0 > 0$

We have seen the defining conditions before in other cases. The effects of these are that $N > 0$ and the other term in the expression for Y is negative. Therefore, Y may be positive, negative, or zero.

In cases 2 through 9, the absolute value of N/M decreases as z increases. A remark on the expression V defined below for the cases where $b_1 \neq 0$ and $b_0 = 0$ is in order. When will $V = 0$? This happens when

$$L_2 = \frac{3E_3 + S^{\frac{1}{2}}}{4E_2}$$ (6.634)

$$L_2 = \frac{3E_3 - S^{\frac{1}{2}}}{4E_2}$$ (6.635)

where

$$S = 9E_3^2 - 8E_2E_4$$ (6.636)

Formulae (6.634) and (6.635) define two critical span lengths L_2, but what is V? To draw our attention in a rather unusual way, we will let it show up later.

For now, let us do something similar to U first. Thus, a similar remark can be made regarding when U will be zero. This happens when

$$L_2 = \frac{E_3 + T^{\frac{1}{2}}}{E_2}$$ (6.636A)

or

$$L_2 = \frac{E_3 - T^{\frac{1}{2}}}{E_2}$$ (6.636B)

where

$$T = E_3^2 - E_2E_4$$ (6.636C)

For the cases where $b_1 \neq 0$ and $b_0 = 0$, the following is offered regarding L_2 as the variable here.

$$b_1 = -\left(\frac{2}{J_3}\right)V$$ (6.636D)

where

$$V = 2L_2^2E_2 - 3L_2E_3 + E_4$$ (6.636E)

Then

$$V' = 4L_2E_2 - 3E_3$$ (6.636F)

$$V'' = 4E_2 > 0$$ (6.636G)

so V has a minimum when

$$V' = 0 \text{ at } L_2 = \frac{3E_3}{4E_2}$$ (6.636H)

6.9.4. Parametric Study of J_1

The parameter is again denoted by z. Here, Z_1 is a constant, while K is given by

$$K = k_0 + \frac{k_1}{z} \tag{6.637}$$

where

$$k_0 = 3C_2^2 + 4C_2C_3, \qquad k_1 = 4(C_2 + C_3)L_1 \tag{6.638}$$

As usual, we consider the function defined by

$$Y = \frac{C}{K} \tag{6.639}$$

where

$$C = \frac{Z_1}{J_2L_1L_2} \tag{6.640}$$

is a constant, with its sign determined by that of Z_1.

We can rewrite Y as

$$Y = \left(\frac{C}{k_0}\right)\left[1 - \frac{N}{z + N}\right] \tag{6.641}$$

where

$$N = \frac{k_1}{k_2} > 0 \tag{6.642}$$

because $k_1 > 0$ and $k_0 > 0$.

Note that $N/(z + N)$ decreases as z increases and there is a minus sign in the bracket in the expression for Y. Therefore, Y is an increasing function of z. Note that Z_1 as well as C may be positive, negative, or zero. The factors of Y in (6.641) other than C are positive, so the sign of C determines that of Y.

6.9.5. Parametric Study of J_2

Denoting the parameter by z, we have

$$K = k_2z^{-2} + k_1z^{-1} + k_0 \tag{6.643}$$

where

$$k_2 = 3L_2^2, \qquad k_1 = 4(C_1 + C_3)L_2, \qquad k_0 = 4C_1C_3 \tag{6.644}$$

and

$$Z_1 = a_1 z^{-1} + a_0 \qquad (6.645)$$

where

$$a_1 = 3L_2 F, \qquad a_0 = 2C_3 G \qquad (6.646)$$

with

$$F = (-L_2^2 E_2 + 2L_2 E_3 - E_4), \qquad G = (-2L_2^2 E_2 + 3L_2 E_3 - E_4) \qquad (6.647)$$

as seen previously on several occasions.

Thus, the function under study is

$$Y = \frac{1}{L_1 L_2} \left(\frac{z Z_1}{P} \right) \qquad (6.648)$$

where

$$P = K z^2 = k_2 + k_1 z + k_0 z^2 \qquad (6.649)$$

Note that $k_2 > 0$, $k_1 > 0$, $k_0 > 0$, $L_1 > 0$, and $L_2 > 0$, whereas a_0 and a_1 may be positive, negative, or zero. We will consider the following cases to get to the bottom of this.

Case 1. $a_0 = 0$

Y is simplified to

$$Y = \frac{a_1}{P L_1 L_2} \qquad (6.650)$$

The sign of Y is determined by that of a_1 because L_1, L_2, and P are all positive, whereas a_1 is positive, negative, or zero when F is positive, negative, or zero respectively. It is clear that the absolute value of Y decreases as z increases.

Let us see what $a_0 = 0$ means by itself. This means that $G = 0$, which leads to

$$-2L_2^2 E_2 + 3L_2 E_3 - E_4 = 0 \qquad (6.651)$$

The roots of this equation, when L_2 is considered as the variable, are

$$L_2 = \frac{3E_3 + Q^{1/2}}{4E_2} \qquad (6.652)$$

$$L_2 = \frac{3E_3 - Q^{1/2}}{4E_2} \qquad (6.653)$$

where

$$Q = 9E_3^2 - 8E_2 E_4 \qquad (6.654)$$

We recognize that (6.652–6.654) are old friends from Section 6.9.3 on the parametric study of L_3. Because of $G = 0$, we have $F = -L_2E_3$ from (6.647). Then, by (6.646), we obtain $a_1 < 0$, meaning $Y < 0$ and $a_1 = -3L_2^2E_3$, a nice result.

Case 2. $a_1 = 0$

The sign of Y is the same as that of a_0. When $a_1 = 0$, we have $F = 0$. This means that

$$-L_2^2E_2 + 2L_2E_3 - E_4 = 0 \qquad (6.655)$$

The roots of this equation, when L_2 is treated as the unknown, are

$$L_2 = \frac{E_3 + S^{1/2}}{E_2} \qquad (6.656)$$

$$L_2 = \frac{E_3 - S^{1/2}}{E_2} \qquad (6.657)$$

where

$$S = E_3^2 - E_2E_4 \qquad (6.658)$$

which, along with (6.656) and (6.657), we met in Section 6.9.3 on L_3.

Case 3. $a_0 \neq 0$ and $a_1 \neq 0$

Next, for the most general case in terms of the values of a_1 and a_0, we consider

$$f = \frac{a_0z + a_1}{P} \qquad (6.659)$$

as a function of z. Then

$$f' = \frac{N}{P^2} \qquad (6.660)$$

where

$$N = (a_0k_2 - a_1k_1) - 2a_1k_0z - a_0k_0z^2 \qquad (6.661)$$

is a second-degree polynomial in z.

The various domains of z that correspond to positive-, negative-, or zero-valued N can be determined by the quadratic formula and a simple procedure which we used on several occasions in previous sections. As always, we could investigate the possibilities of maximum or minimum values of f and thus those of Y with the roots so obtained. Moreover, it appears that, in addition to the roots z_1 and z_2 of $N = 0$, with N being considered as a quadratic equation in z, there is one more way that we can realize $N = 0$ and that is through a choice of all zero coefficients in N. Thus, we may set $a_0 = 0$ and $a_1 = 0$. In this way, we will have $N = 0$ regardless of what value z has. This entails the condition that the span lengths obtained from letting $F = 0$ and $G = 0$ simultaneously are equal. The question remains: Is this possible? That is very interesting, and the reader is encouraged to pursue it.

Note that since $P^2 > 0$, we have $f' < 0, = 0,$ or > 0 when $N < 0, = 0,$ or > 0 respectively. (In fact, $P > 0$ itself.)

Let us now consider the interesting special case where $a_0 k_2 - a_1 k_1 = 0$. Here, N is simplified to

$$N = -k_0 z(2a_1 + a_0 z) \tag{6.662}$$

which leads to the conclusion that $N = 0$ when

$$z = \frac{-2a_1}{a_0} \tag{6.663}$$

provided

$$\frac{a_1}{a_0} < 0 \tag{6.664}$$

with $a_0 \neq 0$.

For (6.664) to hold, a_0 and a_1 need to be of opposite sign. This means that, at least on the surface, either

$$F < 0 \quad \text{and} \quad G > 0 \tag{6.664A}$$

simultaneously or

$$F > 0 \quad \text{and} \quad G < 0 \tag{6.664B}$$

simultaneously.

Expression (6.664A) is possible by resorting to a straightforward proof. What about (6.664B)? It is a wake-up call for the interested reader to find the answer by applying the property of E_j ($j = 2, 3, 4$) established earlier or devising something else that fits.

We just examined the case where $N = 0$ in detail. Let us take care of $N > 0$ now. This happens when

$$a_0 z + 2a_1 < 0 \tag{6.665}$$

That is the case when

$$z < \frac{-2a_1}{a_0} \tag{6.666}$$

which is possible if (6.664) is satisfied, with the same underlying conditions involving F and G properly taken care of.

The case where $N < 0$ can be dealt with in exactly the same manner, with the obvious conclusion that here

$$z > \frac{-2a_1}{a_0} \tag{6.667}$$

along with the same remarks made about the case where $N > 0$.

With all the detailed descriptions about N wrapped up, we will return to the meaning of $a_0 k_2 - a_1 k_1 = 0$ in the present case where a_0 and a_1 are nonzero. We can rewrite this equation as

$$\frac{a_0}{a_1} = \frac{k_1}{k_2} \tag{6.668}$$

which leads to

$$C_3 G = 2F(C_1 + C_3) \tag{6.669}$$

Note that F and G are functions of L_2, E_2, E_3, and E_4 only, whereas (6.669) is a relationship between the set of functions F, G and the beam member properties C_1, C_3. Since F and G each can be positive, negative, or zero and $C_1 > 0$, $C_3 > 0$, the implication of (6.669) is that F and G must be of the same sign. Besides, if F is zero, then G is zero also and vice versa. Furthermore, (6.669) requires that the absolute value of G be greater than that of F.

Note also that N can be rewritten, with the aid of (6.669), as

$$N = -k_0 a_0 z \left[\frac{2k_2}{k_1} + z \right] \tag{6.670}$$

from which we can see clearly that $a_0 < 0$ implies $N > 0$, whereas $a_0 > 0$ implies $N < 0$. Of course, since both z and $(2k_2/k_1 + z)$ are positive for nonzero a_0, which is the present case, we have nonzero N in view of Expression (6.670), which also shows that N is proportional to a_0. Also, remember that $a_0 = 2C_3 G$, where G is given in (6.647).

Note that interesting things happen here: $a_1 = 0$ if and only if $a_0 = 0$ under $a_0 k_2 - a_1 k_1 = 0$, the premise of the special case.

6.9.6. Parametric Study of J_3

The parameter z appears in entities K and Z_1 in the following manner.

$$K = k_0 + k_1 z^{-1} \tag{6.671}$$

where

$$k_0 = 3C_2^2 + 4C_1 C_2, \qquad k_1 = 4(C_1 + C_2)L_3 \tag{6.672}$$

and

$$Z_1 = \frac{a_1}{z} + a_0, \qquad a_1 = 2L_3G, \qquad a_0 = 3C_2F \qquad (6.673)$$

where F and G are as defined in Section 6.9.5.

For the purpose of studying R_1 and R_4, consider the function defined by

$$Y = \left[\frac{1}{J_2 L_1 L_2 k_0} \right] \left[a_0 + \frac{N}{z + \dfrac{k_1}{k_0}} \right] \qquad (6.674)$$

where

$$N = a_1 - \frac{a_0 k_1}{k_0} \qquad (6.675)$$

can be rewritten as

$$N = 2L_3G - 3\left(\frac{k_1}{k_0} \right) C_2F \qquad (6.676)$$

Now $N < 0$, $= 0$, or > 0 when

$$2L_3G - 3\left(\frac{k_1}{k_0} \right) C_2F < 0, = 0, \text{ or } > 0 \qquad (6.677)$$

respectively. Note that C_2, L_3, k_0, and k_1 are all positive.

Case 1. N = 0

This happens when

$$2L_3G = 3\left(\frac{k_1}{k_0} \right) C_2F \qquad (6.678)$$

Here, F and G must have the same sign, and $G = 0$ if and only if $F = 0$.

Note that

$$\frac{k_1}{k_0} = \frac{4(C_1 + C_2)L_3}{(3C_2 + 4C_1) C_2} > 0 \qquad (6.679)$$

Substituting (6.679) into (6.678), we have

$$G = \frac{6F(C_1 + C_2)}{4C_1 + 3C_2} < 6F \frac{C_1 + C_2}{3C_1 + 3C_2} = 2F \qquad (6.680)$$

and

$$G > \frac{6F(C_1 + C_2)}{4C_1 + 4C_2} = 1.5F \tag{6.681}$$

Thus, we see that $2F$ is an upper bound for G, while $1.5F$ is a lower bound for G in the present case.

When $N = 0$, the sign of Y is the same as that of a_0 and

$$Y = \frac{3F}{J_2^2 L_1 k_0} \tag{6.682}$$

Note that here Y is proportional to F, indicating once again the power of F.

In general, we have

$$G = \frac{6F(C_1 + C_2)}{4C_1 + 3C_2} < 0, = 0, \text{ or } > 0 \tag{6.683}$$

corresponding to $N < 0, = 0,$ or > 0 respectively.

The following are three important special cases.

Case 1. $a_0 = 0$

The results are

$$N = a_1 \tag{6.684}$$

$$Y = \left[\frac{1}{J_2 L_1 L_2 k_0} \right] \left[\frac{N}{z + \left(\dfrac{k_1}{k_0} \right)} \right] \tag{6.685}$$

It appears that Y is proportional to N, which is a_1 now with

$$a_1 = 2L_3 G \tag{6.686}$$

where G is reduced to $L_2 E_3$ because $a_0 = 0$. Thus,

$$Y = \left[\frac{1}{J_2 L_1 k_0} \right] \left[\frac{2L_3 E_3}{z + \left(\dfrac{k_1}{k_0} \right)} \right] \tag{6.687}$$

The sign of Y agrees with that of N.

Note that $N = 0$ if and only if $Y = 0$, but with $R_3 = 0$, a trivial load case.

Case 2. $a_1 = 0$

The consequences are

$$N = \frac{-a_0 k_1}{k_0} \tag{6.688}$$

Since N and a_0 have opposite signs, N and F do also.

Note that

$$N = 0 \text{ if and only if } a_0 = 0, \text{ i.e., if and only if } F = 0 \tag{6.689}$$

Thus, $E_3 = 0$, and this is the same as in case 1.

Case 3. $a_0 = a_1 = 0$

This means

$$N = 0 \quad \text{and} \quad Y = 0 \tag{6.690}$$

simultaneously, and

$$G = 0 = F, \quad \text{implying } E_3 = 0 \tag{6.691}$$

We also know the story of these situations from Section 6.9.5.

Let's return to (6.675) and consider Y'. We know that

$$Y' < 0 \text{ for } N > 0 \tag{6.692}$$

$$Y' > 0 \text{ if } N < 0 \tag{6.693}$$

$$Y' = 0 \text{ when } N = 0 \tag{6.694}$$

Thus, N is the key to understanding the behavior of Y.

6.10. TRIANGULAR LOAD ON THE INTERIOR SPAN OF A THREE-SPAN CONTINUOUS BEAM: GENERAL CASE

Following the established procedure, we start with the general expression for the reaction R_1 as

$$R_1 = C[A_0 Z_1 E_2 + (A_0 Z_2 + A_1 Z_1)E_3 + (A_0 Z_3 + A_1 Z_2)E_4 + A_1 Z_3 E_5] \tag{6.695}$$

where

$$C = \frac{L_1 L_3^2}{6D} \tag{6.696}$$

with

$$D = \left(\frac{L_1 L_3}{6}\right)^2 K \qquad (6.697)$$

in which

$$K = 3C_2^2 + 4(C_1 + C_3)C_2 + 4C_1 C_3 \qquad (6.698)$$

The other entities in (6.695) are as follows:

$$A_0 = -2L_2 C_3, \qquad A_1 = -(3C_2 + 2C_3) \qquad (6.699)$$

$$Z_1 = -aU, \qquad Z_2 = \left[1 + \frac{a}{L_2}\right]U, \qquad Z_3 = \frac{-U}{L_2} \qquad (6.700)$$

$$U = \frac{w}{6bJ_2} \qquad (6.701)$$

We notice that these entities exhibit certain attractive features. First of all, C is a ratio between $L_1 L_3^2$ and $6D$. Then, we see that A_0 and A_1 are dependent on L_2, L_3, J_2, and J_3, while Z_1, Z_2, and Z_3 all contain w as a factor through U, which is itself inversely proportional to bJ_2.

For convenience, we rewrite R_1 as

$$R_1 = CP \qquad (6.702)$$

where

$$P = p_2 E_2 + p_3 E_3 + p_4 E_4 + p_5 E_5 \qquad (6.703)$$

with

$$p_2 = A_0 Z_1 = 2aL_2 C_3 U > 0 \qquad (6.704)$$

$$p_3 = A_0 Z_2 + A_1 Z_1 = U(-2L_2 C_3 + 3aC_2) \qquad (6.705)$$

$$p_4 = A_0 Z_3 + A_1 Z_2 = U\left[3C_2\left(1 + \frac{a}{L_2}\right) + \frac{2aC_3}{L_2}\right] < 0 \qquad (6.706)$$

$$p_5 = A_1 Z_3 = \frac{U}{L_2}(3C_2 + 2C_3) > 0 \qquad (6.707)$$

From the last four formulae, we see that each p_i contains U as a common factor and another factor which is specific to that particular p_i.

6.10.1. Parametric Study of L_1

All the p's are constants with respect to the parameter z. Other entities involved in R_1 are functions of z and are described below.

$$K = a_0 + a_1 z \tag{6.708}$$

where

$$a_0 = 3C_2^2 + 4C_2 C_3 > 0 \tag{6.709}$$

$$a_1 = \frac{4(C_2 + C_3)}{J_1} > 0 \tag{6.710}$$

will be used in the expression

$$D = z^2 \left(\frac{L_3}{6} \right)^2 K \tag{6.711}$$

to yield the first factor in R_1 as

$$C = \frac{zL_3^2}{6D} = \frac{6}{z(a_0 + a_1 z)} \tag{6.712}$$

Thus, we have from (6.702) and (6.712)

$$R_1 = \frac{6P}{z(a_0 + a_1 z)} \tag{6.713}$$

where $6P$ is a constant and P is defined in (6.703).

Next, consider the function

$$Y = \frac{1}{z(a_0 + a_1 z)} \tag{6.714}$$

We see that

$$R_1 = 6PY \tag{6.713A}$$

$$Y' = -z^{-2}(a_0 + a_1 z)^{-2} (a_0 + 2a_1 z) < 0 \tag{6.715}$$

because

$$(a_0 + a_1 z)^{-2} > 0, \qquad (a_0 + 2a_1 z) > 0 \tag{6.716}$$

The same remarks as for the case of uniform load apply here.

6.10.2. Parametric Study of L_2

As functions of the parameter z, the pertinent entities are described as follows.

$$D = \left(\frac{L_1 L_3}{6} \right)^2 K \tag{6.717}$$

with

$$K = k_0 + k_1 z + k_2 z^2 \tag{6.718}$$

where

$$k_0 = 4C_1C_3 > 0, \qquad k_1 = \frac{4(C_1 + C_3)}{J_2} > 0, \qquad k_2 = \frac{3}{J_2^2} > 0 \tag{6.719}$$

Thus,

$$C = \frac{L_1 L_3^2}{6D} = \frac{\dfrac{6}{L_1}}{k_0 + k_1 z + k_2 z^2} \tag{6.720}$$

Additionally, we have

$$p_2 = 2aUC_3 z \tag{6.721}$$

$$p_3 = U\left(-2C_3 + \frac{3a}{J_2}\right)z \tag{6.722}$$

$$p_4 = -U\left(\frac{3z}{J_2} + \frac{3a}{J_2} + \frac{2aC_3}{z}\right) \tag{6.723}$$

$$p_5 = U\left(\frac{3}{J_2} + \frac{2C_3}{z}\right) \tag{6.724}$$

The following are some interesting observations. C contains a quadratic expression of z in its denominator, while p_2 and p_3 are proportional to z. As for p_4 and p_5, they both contain the term $1/z$ and a constant term, with the former depending on z also.

Rewriting P as

$$P = \frac{A_2 z^2 + A_1 z + A_0}{z} \tag{6.725}$$

where

$$A_2 = 2aC_3 E_2 + \left(\frac{3a}{J_2} - 2C_3\right)E_3 - \frac{3E_4}{J_2} \tag{6.726}$$

$$A_1 = \frac{3}{J_2}(-aE_4 + E_5) \tag{6.727}$$

$$A_0 = 2C_3(-aE_4 + E_5) \tag{6.728}$$

we note that A_2 is a function of E_j ($j = 2, 3, 4$), a, C_3, and J_2, whereas the coefficient A_1 is a function of E_4, E_5, a, and J_2 and the coefficient A_0 is a function of C_3, a, E_4, and E_5.

Consider the function

$$Y = \frac{L_1 R_1}{6} = \left(\frac{1}{k_2}\right) z \left[A_2 + \frac{N}{z^2 + b_1 z + b_0}\right] \tag{6.729}$$

where

$$N = (A_1 - A_2 b_1)z + (A_0 - A_2 b_0), \qquad b_0 = \frac{k_0}{k_2}, \qquad b_1 = \frac{k_1}{k_2} \tag{6.730}$$

It seems plausible that A_2 is positive, negative, or zero when

$$\frac{A_2}{U} = \left(2aC_3 E_2 + \frac{3aE_3}{J_2}\right) - \left(2C_3 E_3 + \frac{3E_4}{J_2}\right) > 0, < 0, \text{ or } = 0 \tag{6.731}$$

respectively.

Following a similar line of reasoning, it appears natural to state that A_1 is positive, negative, zero when

$$\frac{A_1}{U} = \left(\frac{3}{J_2}\right)(-aE_4 + E_5) > 0, < 0, \text{ or } = 0 \tag{6.732}$$

respectively and A_0 is positive, negative, or zero when

$$\frac{A_0}{U} = 2C_3(-aE_4 + E_5) > 0, < 0, \text{ or } = 0 \tag{6.733}$$

respectively. However, is the situation just described really so straightforward and obvious? Moreover, it is interesting to note that A_1 and A_0 have a common factor which determines their sign.

Recall that E_j ($j = 2, 3, 4, 5$) has some interesting and important attributes which are very useful as well. These are

$$E_5 - aE_4 > 0, \qquad E_4 - aE_3 > 0, \qquad E_3 - aE_2 > 0 \tag{6.734}$$

Therefore, Formula (6.731), upon rearranging the terms, becomes

$$\frac{A_2}{U} = 2C_3(aE_2 - E_3) + \left(\frac{3}{J_2}\right)(aE_3 - E_4) \tag{6.731A}$$

which is negative. This means that A_2 is negative, because both $(aE_2 - E_3)$ and $(aE_3 - E_4)$ are negative, while U, $2C_3$, and $3/J_2$ are all positive.

With these findings, from the expression for N, we see that $N > 0$. When we return to the expression for Y with this information and $A_2 < 0$, we see that Y may be positive, negative, or zero.

As to the effects of varying A_2, A_1, and A_0 on N, we can again see from the expression for N and the sign of A_2, A_1, and A_0 that N increases as the absolute value of these A_i increases. The effects of b_0 and b_1 are similar, even though here the b's are all positive. In order to increase the absolute value of A_2, A_1, and A_0, we can increase the value of one or more of the following: C_3, $1/J_2$, $(E_5 - aE_4)$, $(E_4 - aE_3)$, and $(E_3 - aE_2)$.

Note that a procedure similar to that used in the problem of an exterior span being loaded can be applied here. However, the author chose to use a different approach to raise the reader's interest and enrich the treatment by showing a diversity of approaches.

6.10.3. Parametric Study of L_3

Entities affected by the parameter L_3, denoted as z, are described below.

$$C = \frac{6}{L_1 K} \tag{6.735}$$

where

$$K = k_1 z + k_0 > 0 \tag{6.736}$$

with

$$k_1 = \frac{4(C_1 + C_2)}{J_3} > 0 \tag{6.737}$$

$$k_0 = 3C_2^2 + 4C_1 C_2 > 0 \tag{6.738}$$

Also, we have

$$p_2 = \frac{2aUL_2 z}{J_3} \tag{6.739}$$

$$p_3 = U\left(\frac{-2L_2 z}{J_3} + 3aC_2\right) \tag{6.740}$$

$$p_4 = -U(A_0 + A_1 z), \text{ with } A_0 = 3C_2\left(1 + \frac{a}{L_2}\right), \quad A_1 = \frac{2a}{J_3 L_2} \tag{6.741}$$

$$p_5 = \frac{U}{L_2}\left(3C_2 + \frac{2z}{J_3}\right) \tag{6.742}$$

Note that all the p's are linear functions of z.

Now consider

$$Y = \frac{R_1 L_1}{6} = \left(\frac{U}{k_0 + k_1 z}\right)(B_1 z + B_0) \tag{6.743}$$

where

$$B_1 = \frac{2}{J_2}\left[L_2(aE_2 - E_3) - \left(\frac{1}{L_2}\right)(aE_4 - E_5) \right] \qquad (6.744)$$

$$B_0 = 3\left[aC_2 E_3 - C_2\left(1 + \frac{a}{L_2}\right)E_4 + \frac{E_5}{J_2} \right] \qquad (6.745)$$

Once again, for convenience, we will rewrite Y as

$$Y = \frac{U}{k_1}\left(B_1 + \frac{N}{z + \frac{k_0}{k_1}} \right) \qquad (6.746)$$

where

$$N = B_0 - \frac{B_1 k_0}{k_1} \qquad (6.747)$$

We observe that $N < 0, = 0,$ or > 0 when

$$B_0 - \frac{B_1 k_0}{k_1} < 0, = 0, \text{ or } > 0 \qquad (6.748)$$

respectively. This means that

$$\frac{B_0}{B_1} - \frac{k_0}{k_1} < 0, = 0, \text{ or } > 0 \text{ with nonzero } B_0, B_1 \qquad (6.749)$$

respectively. Note that B_0 as well as B_1 may be positive, negative, or zero. Therefore, we purposefully did not use ratios involving B_0 and B_1 in (6.747) and (6.748).

Case 1. $B_0 = 0$ and $B_1 = 0$

The first condition says that

$$aC_2 E_3 + \frac{E_5}{J_2} = C_2\left(1 + \frac{a}{L_2}\right)E_4 \qquad (6.750)$$

The second condition means that

$$(aE_2 - E_3)L_2^2 = (aE_4 - E_3) \qquad (6.751)$$

which is an extremely interesting result, as will be seen in greater detail in Sections 6.10.5 and 6.10.6 when this special L_3 is called L_{3c}.

The direct result of these two conditions (6.749) and (6.750) together is $N = 0$ and $Y = 0$ simultaneously.

Case 2. $B_0 = 0$

The immediate consequence is $N = -B_1 k_0/k_1$. Here, $N > 0$ or $N < 0$ if and only if $B_1 < 0$ or $B_1 > 0$ respectively.

Additionally, if $B_1 = 0$ also, then it becomes case 1. Thus, B_1 here is nonzero. We have from (6.744)

$$B_1 < 0 \text{ when } L_2(aE_2 - E_3) < \left(\frac{aE_4 - E_5}{L_2}\right) \tag{6.752}$$

$$B_1 > 0 \text{ when } L_2(aE_2 - E_3) > \left(\frac{aE_4 - E_5}{L_2}\right) \tag{6.753}$$

Note that both $aE_2 - E_3$ and $aE_4 - E_5$ are negative except for the trivial case $e = a$, with zero load on the beam.

From the expression for Y, we see that Y is directly proportional to B_1 and the two terms in Y have opposite signs. Therefore, Y may be positive, negative, or zero.

Case 3. $B_1 = 0$

From the expression for N, we see that N is simplified to $N = B_0$. Thus, N is positive or negative when $B_0 > 0$ or $B_0 < 0$ respectively. Here,

$$B_0 > 0 \text{ when } aC_2E_3 + \frac{E_5}{J_2} > C_2\left(1 + \frac{a}{L_2}\right)E_4 \tag{6.754}$$

$$B_0 < 0 \text{ when } aC_2E_3 + \frac{E_5}{J_2} < C_2\left(1 + \frac{a}{L_2}\right)E_4 \tag{6.755}$$

Since Y is proportional to N in this case, Y enjoys the same rights as N does. Moreover, N increases as B_0 increases, and the remarks made earlier about the attributes of B_0 apply here.

Case 4. $B_0 < 0$ and $B_1 > 0$

The immediate implication is $N < 0$. This means

$$\frac{B_0}{B_1} < \frac{k_0}{k_1} \tag{6.756}$$

In the expression for Y, the two terms are of opposite sign. Hence, it is conceivable that Y may be positive, negative, or zero.

Case 5. $B_0 > 0$ and $B_1 < 0$

We have here the direct opposite of case 4 in terms of both the premises and the conclusion regarding the sign of N. However, the conclusion regarding the sign of Y remains the same as in case 4.

We have the case-defining condition

$$B_0 - B_1 \left(\frac{k_0}{k_1} \right) > 0 \tag{6.757}$$

Case 6. $B_0 > 0$ and $B_1 > 0$

In this case, N may be positive, negative, or zero, and thus Y may be positive, negative, or zero. Let us look at the interesting special case where $N = 0$. This means

$$\frac{B_0}{B_1} = \left(\frac{k_0}{k_1} \right) \tag{6.758}$$

and is admissible because B_0 and B_1 have the same sign and the ratio between them is positive, as required by k_0/k_1, which is positive. Now, as a result of $N = 0$, we have

$$Y = \frac{UB_1}{k_1} \tag{6.759}$$

which is independent of z and is proportional to B_1 and inversely proportional to k_1. As to the factor U in the expression for Y, it is there in all cases, and of course, Y is proportional to U without mentioning it in each case.

$N > 0$ when (6.757) holds, and this is the actual situation here. Note that $Y > 0$ when $N > 0$ because both B_1 and N in Y are positive.

For $N < 0$ to hold, the requirement for the sign of B_0 and B_1 is less stringent, and the present case still allows $N < 0$, through (6.756), to happen with the appropriate B_0 and B_1. Here, since $B_1 > 0$ and $N < 0$, it is possible that Y becomes positive, negative, or zero.

Case 7. $B_0 < 0$ and $B_1 < 0$

From the definition of N, we see that N may be positive, negative, or zero. The defining conditions for all these situations are exactly the same as the corresponding ones in case 6, with most of the remarks there applicable here. However, because B_0

< 0 here as opposed to $B_0 > 0$ in case 6, we have the following statement regarding $N < 0$ here: When $N < 0$, we have $Y < 0$ in case 7.

Finally, note that N appears in Y in the form

$$Nf = \frac{N}{z + \dfrac{k_0}{k_1}} \tag{6.760}$$

The function f is a decreasing function of z. Hence, Nf and, for that matter, Y are decreasing functions of z when $N > 0$ and they are increasing functions of z when $N < 0$.

6.10.4. Parametric Study of J_1

Here the entities involved in the reaction R_1 are described below as functions of the parameter $z = J_1$.

$$K = k_0 + \frac{k_1}{z} \tag{6.761}$$

where

$$k_0 = 3C_2^2 + 4C_2C_3, \qquad k_1 = 4(C_2 + C_3)L_1 \tag{6.762}$$

will appear in

$$C = \frac{6}{L_1 K} \tag{6.763}$$

Note that all p_i ($i = 2, 3, 4$) are constants. Hence,

$$R_1 = CG \tag{6.764}$$

where

$$G = p_2E_2 + p_3E_3 + p_4E_4 + p_5E_5 \tag{6.765}$$

is a constant. p_i will be defined below.

We can rewrite R_1 as

$$R_1 = \frac{6G}{L_1 k_0}\left[1 - \frac{N}{z + N}\right] \tag{6.766}$$

where

$$N = \frac{k_1}{k_0} > 0 \tag{6.767}$$

from which we can see that R_1 depends on G, L_1, k_0, k_1, and z. The last four entities just mentioned are all positive. Even though G is a constant, it may be positive, negative, or zero, because in general $p_2 > 0$; p_3 may be positive, negative, or zero; $p_4 < 0$; and $p_5 > 0$. From (6.766), we see also that R_1 is directly proportional to G and inversely proportional to L_1k_0. Moreover, R_1 has an upper bound, namely $6G/(L_1k_0)$.

Now let us go back to G, where the key component is

$$p_3 = U(-2L_2C_3 + 3aC_2) \tag{6.768}$$

The important question is: When will p_3 be positive, negative, or zero and what does each of these situations mean?

$$p_3 < 0, = 0, \text{ or } > 0 \text{ when } 3aC_2 - 2L_2C_3 < 0, = 0, \text{ or } > 0 \tag{6.769}$$

respectively. Each one of these three is a relation among L_2, C_2, a, and C_3. Specifically,

$$p_3 < 0 \text{ when } \frac{3a}{L_2} < \frac{2C_3}{C_2} \tag{6.770}$$

$$p_3 = 0 \text{ when } \frac{3a}{L_2} = \frac{2C_3}{C_2} \tag{6.771}$$

$$p_3 > 0 \text{ when } \frac{3a}{L_2} > \frac{2C_3}{C_2} \tag{6.772}$$

It is interesting to observe that the left-hand side of each of these formulae is the ratio between the loaded length and the length of the loaded span with a numerical multiplier, whereas the right-hand side is twice C_3/C_2. When (6.770) holds, we have $p_3 < 0$, $p_4 < 0$ and $p_2 > 0$, $p_5 > 0$. When (6.771) holds, we have $p_4 < 0$ and $p_2 > 0$, $p_5 > 0$. When (6.772) is valid, then we have all $p_i > 0$ except $p_4 < 0$.

It appears that there are two factors G, H in the reaction R_1 that control the sign of R_1, where

$$H = 1 - \frac{N}{z + N} \tag{6.773}$$

However, it can be shown easily, by noting $N > 0$, $z > 0$, that we have $H > 0$. Therefore, G determines the sign of R_1 in addition to the following.

$R_1 = 0$ when $G = 0$. $G = 0$ means

$$p_2E_2 + p_3E_3 + p_4E_4 + p_5E_5 = 0 \tag{6.774}$$

Note that H is an increasing function of z because $N > 0$ renders $H' > 0$. Additionally, $0 < H < 1$, so $0 < R_1 < 6G/(L_1k_0)$.

Finally, it is interesting to note that when $p_3 = 0$, all the other p's are simplified because of the defining condition for $p_3 = 0$, namely $EL_2 = 3aC_2/(2C_3)$. Thus,

$$p_2 = 3a^2C_2U, \qquad p_4 = -U\left[3C_2 + 2C_3 + \frac{4C_3^2}{3C_2}\right],$$

$$p_5 = \frac{2C_3U}{3aC_2}(3C_2 + 2C_3) \tag{6.775}$$

6.10.5. Parametric Study of J_2

Entities appearing in R_1 take the following forms as functions of the parameter J_2, which is denoted by z.

$$U = \frac{w}{6bz} \tag{6.776}$$

$$K = k_2 z^{-2} + k_1 z^{-1} + k_0 \tag{6.777}$$

where

$$k_2 = 3L_2^2, \qquad k_1 = 4(C_1 + C_3)L_2, \qquad k_0 = 4C_1C_3 \tag{6.778}$$

and

$$C = \frac{6}{L_1 K} \tag{6.779}$$

Thus,

$$R_1 = C(p_2E_2 + p_3E_3 + p_4E_4 + p_5E_5) \tag{6.780}$$

with

$$p_2 = \frac{w}{6bz}(2aL_2C_3) \tag{6.781}$$

$$p_3 = \frac{w}{6bz}\left(-2L_2C_3 + \frac{3aL_2}{z}\right) \tag{6.782}$$

$$p_4 = \frac{-w}{6bz}\left(3L_2\frac{1 + \frac{a}{L_2}}{z} + \frac{2aC_3}{L_2}\right) \tag{6.783}$$

$$p_5 = \frac{w}{6bzL_2}\left(\frac{3L_2}{z} + 2C_3\right) \tag{6.784}$$

Therefore, we can write

$$p_2E_2 + p_3E_3 + p_4E_4 + p_5E_5 = \frac{a_1}{z} + \frac{a_2}{z^2} \tag{6.785}$$

where

$$a_1 = \frac{w}{3b}\left(aL_2E_2 - L_2E_3 - \frac{aE_4}{L_2} + \frac{E_5}{L_2}\right)C_3 \tag{6.786}$$

$$a_2 = \frac{w}{2b}\left[aL_2E_2 - L_2\left(1 + \frac{a}{L_2}\right)E_4 + E_5\right] \tag{6.787}$$

Hence, we have

$$R_1 = \left(\frac{6}{L_1}\right)\frac{a_2 + a_1 z}{k_2 + k_1 z + k_0 z^2} \tag{6.788}$$

Note that $a_1 = 0$ when

$$(aE_2 - E_3)L_2 = \frac{aE_4 - E_5}{L_2} \tag{6.789}$$

and $a_0 = 0$ when

$$(aE_3 - E_4)L_2 = aE_4 - E_5 \tag{6.790}$$

Moreover, R_1 and $a_2 + a_1 a$ have the same sign. For the case where $a_1 = 0$, entities R_1 and a_2 have the same sign, and for the case where $a_2 = 0$, entities R_1 and a_1 have the same sign.

For convenience, let

$$P = k_2 + k_1 z + k_0 z^2 \tag{6.790A}$$

$$T = a_2 + a_1 z \tag{6.790B}$$

Then

$$Y = \frac{T}{P} \tag{6.790C}$$

and

$$Y' = \frac{N}{P^2} \tag{6.790D}$$

where

$$N = (a_1 k_2 - a_2 k_1) - 2a_2 k_0 z - a_1 k_0 z^2 \tag{6.790E}$$

Thus,

$$R_1 = \frac{6Y}{L_1} \tag{6.790F}$$

Note that the same remarks as for the case of uniform load apply here, with a_1 and a_2 playing the roles of a_0 and a_1 there respectively.

It is clear from (6.790A) that we see

$$P > 0 \tag{6.790G}$$

From (6.790B), we know that T may be positive, negative, or zero when

$$a_2 + a_1 z > 0, < 0, \text{ or } = 0 \tag{6.790H}$$

respectively. We observe from (6.790C) and (6.790G) that Y and T have the same sign and also $Y = 0$ if and only if $T = 0$.

From (6.790D), we see that

$$Y' < 0, = 0, \text{ or } > 0 \text{ when } N < 0, = 0, \text{ or } > 0 \tag{6.790I}$$

respectively.

Note that N as shown in (6.790E) is a quadratic equation in the variable z, with coefficients involving a_1, a_2, k_0, k_1, and k_2.

6.10.6. Parametric Study of J_3

Entities entering the expression for R_1 as functions of the parameter J_3 designated as z are displayed below:

$$K = k_0 + \frac{k_1}{z}, \qquad k_0 = 3C_2^2 + 4C_1C_2, \qquad k_1 = 4(C_1 + C_2)L_3 \tag{6.791}$$

$$C = \frac{6}{L_1 K} \tag{6.792}$$

$$p_2 = \frac{2aUL_2L_3}{z} \tag{6.793}$$

$$p_3 = U\left(\frac{-2L_2L_3}{z} + 3aC_2\right) \tag{6.794}$$

$$p_4 = -U\left[3C_2\left(1 + \frac{a}{L_2}\right) + \frac{2aL_3}{L_2 z}\right] \tag{6.795}$$

$$p_5 = \frac{U}{L_2}\left(3C_2 + \frac{2L_3}{z}\right) \tag{6.796}$$

Thus,

$$R_1 = C(p_2 E_2 + p_3 E_3 + p_4 E_4 + p_5 E_5) = \frac{6U}{L_1 k_0} \left[a_0 + \frac{N}{z + \frac{k_1}{k_0}} \right] \qquad (6.797)$$

where

$$a_1 = 2L_2 \left[aL_2 E_2 - L_2 E_3 - \left(\frac{a}{L_2} \right) E_4 + \frac{E_5}{L_2} \right] \qquad (6.798)$$

$$a_0 = 3C_2 \left[aE_3 - \left(1 + \frac{a}{L_2} \right) E_4 + \frac{E_5}{L_2} \right] \qquad (6.799)$$

and

$$N = a_1 - \frac{a_0 k}{k_0} \qquad (6.800)$$

We observe from (6.797) that R_1 is proportional to U and inversely proportional to L_1 and k_0.

Since $U = w/(bJ_2) > 0$, $L_1 > 0$, $k_0 > 0$, and $k_1 > 0$, we know that a_1, a_0, and N can be positive, negative, or zero and will look into individual cases. A note is in order before we embark on this undertaking: $E_k - aE_j > 0$ ($j = 2, 3, 4; k = 3, 4, 5$). This is important because expressions of this kind appear quite often and we need to emphasize this point right now.

Category A. $a_1 = 0$

This happens when $L_2 = L_{c1}$, where L_{c1} is defined through

$$L_2^2 = \frac{aE_4 - E_5}{aE_2 - E_3} > 0 \qquad (6.801)$$

Because of its premier importance and frequent occurrence, we will call

$$L_{c1} = \left(\frac{aE_4 - E_5}{aE_2 - E_3} \right)^{1/2} \qquad (6.801A)$$

the critical span length of the first kind.

We note that for this category, N and a_0 have opposite signs. Also, we observe that both N and R_1 are proportional to a_0, while N itself is proportional to k_1/k_0.

Note that the span length to be defined by (6.802) below is not only beautiful and interesting but also very useful, and we will call it the critical span length, L_{c2}, of the second kind, with the property $a_0 < 0$, $= 0$, or > 0 when $L_{c2} < L_2$, $= L_2$, or $> L_2$ respectively.

Case 1. $a_0 = 0$. This happens when $L_2 = L_{c2}$, where L_{c2} is defined by

$$L_{c2} = \frac{aE_4 - E_5}{aE_3 - E_4} \tag{6.802}$$

Some consequences of $a_0 = 0$ and $a_1 = 0$ are $N = 0$ and $R_1 = 0$.

Case 2. $a_0 < 0$. This happens when

$$aE_3 + \frac{E_5}{L_2} < \left(1 + \frac{a}{L_2}\right)E_4 \tag{6.803}$$

with the consequence that $N > 0$. Therefore, R_1 may be positive, negative, or zero because the two terms of R_1 are of opposite signs.

Case 3. $a_0 > 0$. The defining condition is

$$aE_3 + \frac{E_5}{L_2} > \left(1 + \frac{a}{L_2}\right)E_4 \tag{6.804}$$

with the effect that N is rendered negative. Once again, R_1 may be positive, negative, or zero for the same reason as in the last case.

Category B. $a_1 < 0$

This happens when

$$L_2^2 > \frac{aE_4 - E_5}{aE_2 - E_3} \tag{6.805}$$

Case 1. $a_0 = 0$. The defining condition is (6.794), with the result that $N = a_1 < 0$ and $R_1 < 0$. Moreover, R_1 is proportional to N, which is a_1. We see from the definition of a_1 that R_1 here is proportional to L_3 also.

Case 2. $a_0 < 0$. The defining condition is (6.803). The consequences are as follows. First, N may be positive, negative, or zero when $a_1 - a_0 k_1/k_0 > 0$, < 0, or $= 0$ respectively. Second, R_1 may be positive, negative, or zero.

A quick check is provided here for the first conclusion. $a_1/a_0 > 0$ because a_0 and a_1 have the same sign. Given $k_1/k_0 > 0$, it is possible that one positive quantity is less than, greater than, or equal to the other.

Case 3. $a_0 > 0$. The defining condition is (6.804) or equivalently $L_{c2} > L_2$. In this case, $N < 0$, but R_1 may be positive, negative, or zero because the two terms in the expression for R_1 are of opposite signs.

Category C. $a_1 > 0$
The defining condition is

$$L_2^2 < \frac{aE_4 - E_5}{aE_2 - E_3} \tag{6.806}$$

Case 1. $a_0 = 0$. This is the case where $L_{c2} = L_2$, and the results are $N = a_1 > 0$, with R_1 proportional to N and thus also to a_1.

Case 2. $a_0 < 0$. The defining condition is $L_{c2} < L_2$, with the result that $N > 0$ and $R_1 < 0$, $= 0$, or > 0 is possible.

Case 3. $a_0 > 0$. The defining condition is $L_{c2} > L_2$. Here, $N < 0$, $= 0$, or > 0 when

$$a_1 - \frac{a_0 k_1}{k_0} < 0, \; = 0, \; \text{or} \; > 0 \tag{6.807}$$

respectively. Since $k_1/k_0 > 0$ and $a_1/a_0 > 0$, we see that all the situations are admissible.

Note that in all cases where $N = 0$, z has no effect on R_1. Also, with nonzero N, the reaction R_1 is a decreasing or increasing function of z when $N > 0$ or $N < 0$ respectively.

Table 6.3 is a summary table showing all the different cases. There are three possible ways to render $N = 0$ and five possible ways to realize $N < 0$, with yet another five ways to achieve $N > 0$.

6.11. A CONCENTRATED FORCE AT AN ARBITRARY POINT ON AN EXTERIOR SPAN OF A THREE-SPAN CONTINUOUS BEAM: CONSTANT J
The reaction R_1 is given by

$$R_1 = (As + BK)\frac{P}{D} \tag{6.808}$$

where

$$A = L_1 L_2 L_3^2 \left[\frac{L_2 + L_3}{18J^2} \right] = \frac{a_1}{J^2} \tag{6.809}$$

TABLE 6.3. Classification of Cases

	a_0	a_1	
$N = 0$	0	0	
	< 0	< 0	and $a_1/a_0 = k_1/k_0$
	> 0	> 0	and $a_1/a_0 = k_1/k_0$
$N < 0$	> 0	0	
	0	< 0	
	> 0	< 0	
	< 0	< 0	and $a_1/a_0 < k_1/k_0$
	> 0	> 0	and $a_1/a_0 < k_1/k_0$
$N > 0$	< 0	0	
	< 0	< 0	and $a_1/a_0 > k_1/k_0$
	0	> 0	
	< 0	> 0	
	> 0	> 0	and $a_1/a_0 > k_1/k_0$

$$B = \frac{L_1 L_2 L_3}{6J} \tag{6.810}$$

$$K = \frac{s}{6J} Q \tag{6.811}$$

with

$$Q = (s^2 - 3L_3 s - 2L_2 L_3) \tag{6.812}$$

in (6.811) for convenience in future use, and finally

$$D = \frac{d}{J^2} \tag{6.813}$$

with

$$d = \left(\frac{L_1 L_3}{6}\right)^2 F, \quad F = 3L_2^2 + 4(L_1 + L_3)L_2 + 4L_1 L_3 \tag{6.814}$$

Thus,

$$R_1 = (a_1 s + a_2) \frac{P}{d} \tag{6.815}$$

where

$$a_1 = L_1 L_2 L_3^2 \frac{L_2 + L_3}{18} \tag{6.816}$$

$$a_2 = \frac{L_1 L_2 L_3 s Q}{36} \tag{6.817}$$

Note that from (6.814–6.817) R_1 is free of J, and this is very important. Also note that a_2 may be positive, negative, or zero depending on whether $sQ > 0$, < 0, or $= 0$ respectively.

However, it is given that $a_1 > 0$. As a result, the sign of $(a_1 s + a_2)$ depends on the relative magnitude of a_1 and a_2 in any particular situation. Therefore, it is conceivable that R_1 need not be positive all the time.

Now let us proceed to the study of R_4 expressed as

$$R_4 = (Us - GK) \frac{P}{D} \tag{6.818}$$

where

$$U = -\left(\frac{L_1 L_2}{6J}\right)^2 L_3 \tag{6.819}$$

$$G = \frac{(L_1 + L_2)L_1^2}{3J} \tag{6.820}$$

D and K are as defined earlier in this section.

Thus,

$$Us - GK = (b_1 - b_2) \frac{s}{J^2} \tag{6.821}$$

where

$$b_1 = \frac{-L_1^2 L_2^2 L_3}{36} < 0 \tag{6.822}$$

$$b_2 = \frac{(L_1 + L_2)L_1^2 Q}{18} \tag{6.823}$$

and R_4 is given by

$$R_4 = (b_1 - b_2) \frac{sP}{d} \tag{6.824}$$

We note from (6.824) that R is free of J also. Note that $b_1 < 0$, but b_2 may be positive, negative, or zero depending on whether $Q > 0$, $= 0$, or < 0 respectively. However, it is clear from (6.824) and (6.823) that $Q > 0$ will give us $R_4 < 0$. Only when Q is negative and has a large enough absolute value will it be possible to render R_4 zero or positive via b_1 and b_2 alone. Why did we say the last several words in the last sentence? Well, it is interesting to note that R_4 has an outstanding factor, namely s, which can make R_4 vanish along with it.

6.12. A CONCENTRATED FORCE AT AN ARBITRARY POINT ON AN EXTERIOR SPAN OF A THREE-SPAN CONTINUOUS BEAM: CONSTANT SPAN LENGTH

Pertinent entities involved in the expression for $R_1 = P(As + BK)/D$ are described as follows.

$$A = L^5 \left(\frac{\dfrac{1}{J_2} + \dfrac{1}{J_3}}{18 J_2} \right) \tag{6.825}$$

$$B = \frac{L^3}{6 J_2} \tag{6.826}$$

$$D = L^6 d \tag{6.827}$$

where

$$d = \frac{1}{36} \left[\frac{3}{J_2^2} + 4 \left(\frac{1}{J_1} + \frac{1}{J_3} \right) \frac{1}{J_2} + \frac{4}{J_1 J_3} \right] \tag{6.828}$$

$$K = s \, \frac{\left(\dfrac{s^2}{J_3} - \dfrac{3Ls}{J_3} - \dfrac{2L^2}{J_2} \right)}{6} \tag{6.829}$$

Thus

$$R_1 = s \left(\frac{a_1}{L} + \frac{a_2}{L^3} \right) \frac{P}{d} \tag{6.830}$$

Let $x = 1/L$ and define

$$Y = a_1 x + a_2 x^3 \tag{6.831}$$

where

$$a_1 = \frac{\dfrac{1}{J_2} + \dfrac{1}{J_3}}{18 J_2} \tag{6.831A}$$

$$a_2 = \frac{K}{36 J_2} \tag{6.831B}$$

We have

$$R_1 = \frac{PsY}{d} \tag{6.831C}$$

and

$$Y' = a_1 + 3a_2x^2, \qquad Y'' = 6a_2x \qquad (6.832)$$

as well as

$$Y'' = 0, < 0, \text{ or } > 0 \text{ when } a_2 = 0, < 0, \text{ or } > 0 \qquad (6.833)$$

respectively.

For $y' = 0$, we have

$$x^2 = \frac{-a_1}{3a_2} > 0 \qquad (6.834)$$

However, $a_1 > 0$, so (6.834) is possible when $a_2 < 0$. Here, $Y'' = 6a_2x < 0$ and we have a maximum for Y.

When $a_2 < 0$ we have, from (6.831B), $K < 0$. This means that

$$\frac{s^2}{J_3} - \frac{3Ls}{J_3} - \frac{2L^2}{J_2} < 0 \qquad (6.835)$$

Consider the special case where $a_2 = 0$ corresponds to $K = 0$ and $Y' = a_1 > 0$, with $Y'' = 0$. Moreover, Y is positive and proportional to a_1 and x, while R_1 is positive also and proportional to a_1, s, and x. Note that the key thing here is the entity K. When will K be zero? The answer is when the span length is equal to

$$L = \frac{\dfrac{-3s}{J_3} + S^{\frac{1}{2}}}{\dfrac{4}{J_2}} \qquad (6.836)$$

where

$$S = \frac{9s^2}{J_3^2} + \frac{8s^2}{J_2 J_3} \qquad (6.837)$$

The span length shown in (6.836) is a critical span length, as we have just witnessed.

We see that $R_1 = 0, < 0,$ or > 0 when $Y = 0, < 0,$ or > 0 respectively. For nonzero a_2, let us ask: When will $Y = 0$? The answer is

$$Y = 0 \text{ when } a_1 + a_2x^2 = 0 \qquad (6.831D)$$

that is, when

$$x^2 = \frac{-a_1}{a_2} > 0 \qquad (6.831E)$$

where a_1 and a_2 must be of opposite signs. Since $a_1 > 0$ by definition we are left with $a_2 < 0$, which means, by (6.831B),

$$K < 0 \qquad (6.831F)$$

Furthermore,

$$Y > 0 \text{ when } x^2 > \frac{-a_1}{a_2} > 0 \tag{6.831G}$$

$$Y < 0 \text{ when } x^2 < \frac{-a_1}{a_2} > 0 \tag{6.831H}$$

Let us do R_4 now. R_4 is given by

$$R_4 = (Us - GK)\frac{P}{D} \tag{6.838}$$

where

$$U = L^5(36J_2^2) \tag{6.839}$$

$$G = \left(\frac{1}{J_1} + \frac{1}{J_2}\right)\frac{L^2}{3} \tag{6.840}$$

and K as well as P are as shown earlier in this section.

Thus,

$$R_4 = s\left(\frac{b_1}{L} - \frac{b_2}{L^3}\right)\frac{P}{D} \tag{6.841}$$

where

$$b_1 = \frac{-1}{36J_2^2} < 0 \tag{6.842}$$

$$b_2 = \left(\frac{1}{J_1} + \frac{1}{J_2}\right)\frac{K}{18} \tag{6.843}$$

Note that $b_2 < 0$, = 0, or > 0 when $K < 0$, = 0, or > 0 respectively. We realize the power of K again.

Let $1/L = x$ and consider the function

$$X = b_1 x - b_2 x^3 \tag{6.844}$$

Then

$$R_4 = \frac{sX}{d} \tag{6.845}$$

Consider $X' = b_1 - 3b_2 x^2$ and $x'' = -6b_2 x$. We have $X' = 0$ when

$$x^2 = \frac{b_1}{3b_2} \tag{6.846}$$

is positive, for which we need $b_2 < 0$. This means that $K < 0$ and $X'' > 0$. Thus, the x value in (6.846) corresponds to a minimum X.

The special case where $b_2 = 0$ means that $X' = b_1 < 0$, $X'' = 0$, $K = 0$, $X < 0$, and R_4 is negative and proportional to sxb_1.

The special case where $b_2 > 0$ means that $X' < 0$, $X'' < 0$, $X < 0$, $R_4 < 0$, and $K > 0$. Note that in all cases, R_4 has a factor sx, which means that R_4 is always proportional to sx.

For nonzero b_2, let us see when R_4 will be zero or positive or negative. First, we notice that from (6.845) R_4 and X have the same sign; one is equal to zero when the other is. From (6.844), $X = 0$ when

$$x = \left(\frac{b_1}{b_2} \right)^{1/2} \tag{6.847}$$

where $b_2 < 0$ is needed for x in (6.847) to be a real number, because $b_1 < 0$, and the use of $b_2 < 0$ means $K < 0$, of course. From (6.844), again we have $X > 0$ when

$$b_1 > b_2 x^2 \quad \text{or} \quad x < \left(\frac{b_1}{b_2} \right)^{1/2} \tag{6.848}$$

Similarly, we obtain $X < 0$ when

$$b_1 < b_2 x^2 \quad \text{or} \quad x > \left(\frac{b_1}{b_2} \right)^{1/2} \tag{6.849}$$

In both (6.848) and (6.849), $b_2 < 0$ is required because $b_1 < 0$ and we need b_1 and b_2 to have the same sign in order to obtain $b_1/b_2 > 0$ called for in (6.848) and (6.849), just as in (6.847).

6.13. UNIFORM LOAD ON AN EXTERIOR SPAN OF A THREE-SPAN CONTINUOUS BEAM: CONSTANT *J*

The reaction R_1 is given as

$$R_1 = \frac{h_1 E_2 + h_2 E_3 + h_3 E_4}{D} \tag{6.850}$$

where

$$h_1 = \frac{L_1 L_2 L_3^3}{18 J^2} \tag{6.851}$$

$$h_2 = \frac{-L_1 L_2 L_3^2}{12 J^2} \tag{6.852}$$

$$h_3 = \frac{L_1 L_2 L_3}{36 J^2} \qquad (6.853)$$

$$D = \frac{d}{J^2} \qquad (6.854)$$

$$d = \left[\frac{L_1 L_3}{6}\right]^2 [3L_2^2 + 4(L_1 + L_2)L_2 + 4L_1 L_3] \qquad (6.855)$$

Thus, we have

$$R_1 = \left[\frac{L_1 L_2 L_3}{d}\right]\left(\frac{L_3^2 E_2}{18} - \frac{L_3 E_3}{12} + \frac{E_4}{36}\right) \qquad (6.856)$$

which is independent of J.

Note that $R_1 = 0$ when

$$\frac{L_3^2 E_2}{18} - \frac{L_3 E_3}{12} + \frac{E_4}{36} = 0 \qquad (6.857)$$

theoretically. But is it possible for (6.857) to have physical meaning? We will solve (6.857) as a quadratic equation of L_3. Thus,

$$L_{31} = \frac{3E_3 + S^{1/2}}{4E_2} \qquad (6.858)$$

$$L_{32} = \frac{3E_3 - S^{1/2}}{4E_2} \qquad (6.859)$$

where

$$S = 9E_3^2 - 8E_2 E_4 \qquad (6.860)$$

It can be shown easily, from the definitions of E_2, E_3, and E_4, that $S > 0$ in general and $S = 0$ only when $a = e$ or $a = 0$. Therefore, we have two real roots of (6.857) and both of them are positive, because $S < 9E_3^2$. Thus, we have two span lengths L_{31} and L_{32} given by (6.858) and (6.859) at which $R_1 = 0$. Moreover, when $L_3 < L_{32}$ or $L_3 > L_{31}$, we have $R_1 > 0$, while $L_{32} < L_3$ yields $R_1 < 0$. We can even examine a factor of R_1 as a function of L_3, henceforth called Y for convenience, a step further.

Let

$$R_1 = Y\left[\frac{L_1 L_2 L_3}{36 d}\right] \qquad (6.861)$$

$$Y = a_1 x^2 + a_2 x + a_3 \qquad (6.862)$$

where

$$a_1 = 2E_2, \qquad a_2 = -3E_3, \qquad a_3 = E_4 \qquad (6.863)$$

Thus,

$$Y' = 2a_1 x + a_2 \qquad (6.864)$$

For $Y' = 0$, we have

$$x = \frac{3E_3}{4E_2} \qquad (6.865)$$

At this point, $Y'' = 2a_1 > 0$, so Y has a minimum here. It is interesting to note that x has an upper bound $(e + a)/2$.

Let us do R_4 now. We have, from Section 3.2.3 on the general solution,

$$R_4 = \frac{k_1 E_2 + k_2 E_3 + k_3 E_4}{D} \qquad (6.866)$$

where

$$k_1 - L_1^2 L_2 L_3 (4L_1 + 3L_2) \frac{1}{36J^2} \qquad (6.867)$$

$$k_2 = (L_1 + L_2) \frac{L_1^2 L_3}{6J^2} \qquad (6.868)$$

$$k_3 = -(L_1 + L_2) \frac{L_1^2}{18J^2} \qquad (6.869)$$

Thus,

$$R_4 = \frac{L_1^2}{36d} [L_2 L_3 (4L_1 + 3L_2) E_2 + 6(L_1 + L_2) L_3 E_3 \\ - 2(L_1 + L_2) E_4] \qquad (6.870)$$

Note that R_4 is independent of J also.

From (6.870), we see also that R_4 may be positive, negative, or zero when the expression in brackets in (6.870) is positive, negative, or zero respectively.

Also from (6.870), we see that R_4 has an upper bound R_U and a lower bound R_L

$$R_U = \left(\frac{L_1^2}{36d} \right) P_U \qquad (6.871)$$

$$R_L = \left(\frac{L_1^2}{36d} \right) P_L \qquad (6.872)$$

where

$$P_U = (L_1 + L_2)[4L_2L_3E_2 + 6L_3E_3 - 2E_4] \tag{6.873}$$

$$P_L = (L_1 + L_2)[3L_2L_3E_2 + 6L_3E_3 - 2E_4] \tag{6.874}$$

Note that R_U may be positive, negative, or zero when P_U is positive, negative, or zero respectively.

Also,

$$R_L > 0, < 0, \text{ or } = 0 \tag{6.875}$$

when

$$P_L > 0, < 0, \text{ or } = 0 \tag{6.876}$$

respectively.

6.14. UNIFORM LOAD ON AN EXTERIOR SPAN OF A THREE-SPAN CONTINUOUS BEAM: CONSTANT SPAN LENGTH

The reaction R_1 is given by

$$R_1 = \frac{h_1 E_2 + h_2 E_3 + h_3 E_4}{D} \tag{6.877}$$

where

$$h_1 = \frac{L^5}{18 J_2 J_3} \tag{6.878}$$

$$h_2 = \frac{-L^4}{12 J_2 J_3} \tag{6.879}$$

$$h_3 = \frac{L^3}{36 J_2 J_3} \tag{6.880}$$

Thus,

$$R_1 = \frac{Y}{36 d J_2 J_3} \tag{6.881}$$

where

$$Y = 2E_2 x - 3E_3 x^2 + E_4 x^3 \tag{6.882}$$

with

$$x = \frac{1}{L} \tag{6.883}$$

Then,

$$Y' = E_2 - 6E_3x + 3E_4x^2 \tag{6.884}$$

Thus, $Y' = 0$ when

$$x_1 = \frac{3E_3 + S^{1/2}}{3E_4} \tag{6.885}$$

or

$$x_2 = \frac{3E_3 - S^{1/2}}{3E_4} \tag{6.886}$$

where

$$S = 9E_3^2 - 6E_2E_4 \tag{6.887}$$

must be nonnegative in order to have real roots. This means that

$$3E_3^2 - 2E_2E_4 \geq 0 \tag{6.887A}$$

which is shown to be true in Appendix B. We have two positive real roots of $Y' = 0$ given by (6.885) and (6.886), with $x_2 < x_1$. Note that $Y' > 0$ when $x > x_1$ or $x < x_2$, while $Y' < 0$ if $x_2 < x < x_1$.

Let us check $Y'' = -6E_3 + 6E_4$. We have $Y'' < 0$ at $x = x_2$, meaning a maximum Y here, but $Y'' > 0$ at $x = x_1$, signifying a minimum Y at the location given by (6.885).

Next, let us look at Y itself. Can $Y = 0$? Yes, at least $Y = 0$ at $x = 0$, which is a trivial case meaning infinitely long spans. The other values of x at which $Y = 0$ are obtained via

$$2E_2 - 3E_3x + E_4x^2 = 0 \tag{6.888}$$

as

$$x_3 = \frac{3E_3 + T^{1/2}}{2E_4} \tag{6.889}$$

$$x_4 = \frac{3E_3 - T^{1/2}}{2E_4} \tag{6.890}$$

where

$$T = 9E_3^2 - 8E_2E_4 \tag{6.891}$$

The interested reader is encouraged to assess the admissibility of (6.889) and (6.890) as real roots of (6.888) using the strategies employed in solving the corresponding problem associated with $Y' = 0$ or some other method.

Next, we do R_4, which is given by

$$R_4 = \frac{k_1 E_2 + k_2 E_3 + k_3 E_4}{D} \tag{6.892}$$

where D is as in the case for R_1 in this section and

$$k_1 = \frac{L^5\left(\dfrac{4}{J_1} + \dfrac{3}{J_2}\right)}{36 J_2} \tag{6.893}$$

$$k_2 = \frac{L^4\left(\dfrac{1}{J_1} + \dfrac{1}{J_2}\right)}{6 J_3} \tag{6.894}$$

$$k_3 = \frac{-L^3\left(\dfrac{1}{J_1} + \dfrac{1}{J_2}\right)}{18 J_3} \tag{6.895}$$

Thus,

$$R_4 = \frac{Y}{36d} \tag{6.896}$$

where

$$Y = a_1 x + a_2 x^2 + a_3 x^3, \qquad x = \frac{1}{L} \tag{6.897}$$

with

$$a_1 = \frac{\left(\dfrac{4}{J_1} + \dfrac{3}{J_2}\right)E_2}{J_2} > 0 \tag{6.898}$$

$$a_2 = 6\,\frac{\left(\dfrac{1}{J_1} + \dfrac{1}{J_2}\right)E_3}{J_3} > 0 \tag{6.899}$$

$$a_3 = \frac{-2\left(\dfrac{1}{J_1} + \dfrac{1}{J_2}\right)E_4}{J_3} < 0 \tag{6.900}$$

Now,

$$Y' = a_1 + 2a_2 x + 3a_3 x^2 \tag{6.901}$$

so $Y' = 0$ when

$$x_1 = \frac{-a_2 + S^{1/2}}{3a_3} \tag{6.902}$$

$$x_2 = \frac{-a_2 - S^{1/2}}{3a_3} \tag{6.903}$$

where

$$S = a_2^2 - 3a_1 a_3 \tag{6.904}$$

Note that even though $S > 0$, thereby guaranteeing real roots of $Y' = 0$, only one root is positive and, therefore, admissible. This root is given by (6.903). Also, $Y'' = 2a_2 + 6a_3 x < 0$ at x_2, so Y has a maximum there. Additionally, $Y' < 0$ when $x > x_2$, and $Y' > 0$ if $x < x_2$.

Let us deal with $Y = 0$ now. This mean that $R_4 = 0$ also and happens when $x = 0$ or $V = a_1 + a_2 x + a_3 x^2 = 0$. $V = 0$ means

$$x_1 = \frac{-a_2 + T^{1/2}}{2a_3} \tag{6.905}$$

or

$$x_2 = \frac{-a_2 - T^{1/2}}{2a_3} \tag{6.906}$$

where

$$T = a_2^2 - 4a_1 a_3 \tag{6.907}$$

is greater than zero, because $a_1 a_3 < 0$ and thus $-4a_1 a_3 > 0$. However, only one root, given by (6.906), is positive. Again, $Y < 0$ when $x > x_2$ and $Y > 0$ if $x < x_2$. Of course, $R_4 > 0$ when $Y > 0$ and vice versa, as can be seen from (6.896).

It is also very interesting to note in passing that Y and, for that matter, R_4 have a rather beautiful upper bound. First of all, we see $0 < a_1 < 4(1/J_1 + 1/J_2)E_2/J_2$. Then we observe that a_2 and a_3 have a common factor $(1/J_1 + 1/J_2)$. Thus, we have

$$Y < 2\left(\frac{1}{J_1} + \frac{1}{J_2}\right)\left[\frac{2E_2}{J_2} + \left(\frac{3E_3}{J_3}\right)x - \left(\frac{E_4}{J_3}\right)x^2\right]x = B \tag{6.908}$$

and

$$R_4 < \frac{B}{36d} \tag{6.909}$$

Here, B stands for "beautiful," and we have seen something like it, at least in mathematical form, before and maybe on several occasions. Additionally, $Y'' = B''$. When we have a maximum for Y at x, we have also a maximum for B in the neighborhood and similarly for a minimum. The domains of increasing function behavior for B and Y are the same. The same statement can be made regarding decreasing function characteristics.

6.15. TRIANGULAR LOAD ON AN EXTERIOR SPAN OF A THREE-SPAN CONTINUOUS BEAM: CONSTANT J

The reaction R_1 is given by

$$R_1 = \left(\frac{Q}{D}\right)F \tag{6.910}$$

where

$$Q = \frac{L_1 L_2}{6z}, \qquad D = \frac{d}{z^2} \tag{6.911}$$

$$F = (f_1 E_2 + f_2 E_3 + f_3 E_4 + f_4 E_5) \tag{6.912}$$

with

$$d = \left[\frac{L_1 L_3}{6}\right]^2 [3L_2^2 + 4(L_1 + L_3)L_2 + 4L_1 L_3] \tag{6.913}$$

and

$$f_1 = \frac{\dfrac{L_3^3 q_0}{3}}{z} = \frac{a_1}{z} \tag{6.914}$$

$$f_2 = \frac{\dfrac{-L_3^2 q_0}{2} + \dfrac{L_3^3 q_1}{3}}{z} = \frac{a_2}{z} \tag{6.915}$$

$$f_3 = \frac{\dfrac{L_3 q_0}{6} - \dfrac{L_3^2 q_1}{2}}{z} = \frac{a_3}{z} \tag{6.916}$$

$$f_4 = \frac{\dfrac{L_3 q_1}{6}}{z} = \frac{a_4}{z} \tag{6.917}$$

As a result of substituting these entities involved into (6.910), we have

$$R_1 = BH \qquad (6.918)$$

where

$$B = \frac{L_1 L_2}{6d}, \qquad H = a_1 E_2 + a_2 E_3 + a_3 E_4 + a_4 E_5 \qquad (6.919)$$

We observe that B and H are free of z; R_1 is the product of B and H. The conclusions of these simple observations are that R_1 is independent of z and is proportional to B and H. Since B is proportional to both L_1 and L_2 and is inversely proportional to d, so is R_1. Note, of course, that the L's are in d also and they are not independent of each other; the last statement is for figurative comparison purposes only.

From (6.918) and (6.919), we see that if $H < 0$, $= 0$, or > 0, then $R_1 < 0$, $= 0$, or > 0 respectively. Note that H is proportional to L_3 because all the a_i's have this factor in common.

Next, we will study R_4, expressed as

$$R_4 = \frac{G}{D} \qquad (6.920)$$

where

$$G = (g_1 E_2 + g_2 E_3 + g_3 E_4 + g_4 E_5) \qquad (6.921)$$

with

$$g_1 = \frac{d_1 q_0}{z^2} = \frac{b_1}{z^2} \qquad (6.922)$$

$$g_2 = \frac{d_2 q_0 + d_1 q_1}{z^2} = \frac{b_2}{z^2} \qquad (6.923)$$

$$g_3 = \frac{d_3 q_0 + d_2 q_1}{z^2} = \frac{b_3}{z^2} \qquad (6.924)$$

$$g_4 = \frac{d_3 q_1}{z^2} = \frac{b_4}{z^2} \qquad (6.925)$$

and

$$d_1 = -\left(\frac{L_1^2 L_2 L_3}{36}\right)(4L_1 + 3L_2) < 0 \qquad (6.926)$$

$$d_2 = \frac{-L_1^2 L_3 (L_1 + L_2)}{6} < 0 \qquad (6.927)$$

$$d_3 = \frac{(L_1 + L_2)L_1^2}{18} > 0 \tag{6.928}$$

Thus,

$$R_4 = \frac{b_1E_2 + b_2E_3 + b_3E_4 + b_4E_5}{d} \tag{6.929}$$

We observe that all the terms in the numerator of (6.929) have the factor L_1 in common and d has a factor L_1, so they cancel each other out, and we have a simplification there when we carry out the calculations. Note again that R_4 is independent of z.

An examination of the above formulae reveals that $R_4 < 0, = 0,$ or > 0 if $G < 0, = 0$ or > 0 respectively.

6.16. TRIANGULAR LOAD ON AN EXTERIOR SPAN OF A THREE-SPAN CONTINUOUS BEAM: CONSTANT SPAN LENGTH

Pertinent entities involved in the expression for reactions R_1 and R_4 are expressed as functions of the parameter designated as z below.

$$G = \left(\frac{1}{J_1} + \frac{1}{J_2} \right) \frac{z^3}{3} \tag{6.930}$$

$$D = (L^6 d) \frac{3B_2^2 + 4(B_1 + B_3)B_2 + 4B_1B_3}{36} \tag{6.931}$$

$$B_i = \frac{1}{J_i}, \quad i = 1, 2, 3 \tag{6.931A}$$

$$Q = \frac{z^2}{6J_2} \tag{6.932}$$

$$A_1 = \frac{z^3}{3J_2}, \qquad A_2 = \frac{z^2}{2J_3}, \qquad A_3 = \frac{z}{6J_3} \tag{6.933}$$

$$D_1 = d_1 z^5 = \left[-\frac{\dfrac{4}{J_1} + \dfrac{3}{J_2}}{36J_2} \right] z^5 \tag{6.934}$$

$$D_2 = d_2 z^4 = -\left[\frac{\dfrac{1}{J_1} + \dfrac{1}{J_2}}{6J_3}\right] z^4 \tag{6.935}$$

$$D_3 = d_3 z^3 = \left[\frac{\dfrac{1}{J_1} + \dfrac{1}{J_2}}{18J_3}\right] z^3 \tag{6.936}$$

Thus, we have the elements for R_1 as

$$f_1 = A_1 q_0 = \frac{q_0}{3J_3} z^3 \tag{6.937}$$

$$f_2 = A_2 q_0 + A_1 q_1 = \left[\frac{q_1}{3J_3}\right] z^3 - \left[\frac{q_0}{2J_3}\right] z^2 \tag{6.938}$$

$$f_3 = A_3 q_0 + A_2 q_1 = \left[\frac{\dfrac{-q_1}{2}}{J_3}\right] z^2 + \frac{q_0}{6J_3} z \tag{6.939}$$

$$f_4 = A_3 q_1 = \frac{q_1}{6J_3} z \tag{6.940}$$

and the elements for R_4 as

$$g_1 = D_1 q_0 = d_1 q_0 z^5 \tag{6.941}$$

$$g_2 = D_2 q_0 + D_1 q_1 = d_2 q_0 z^4 + d_1 q_1 z^5 \tag{6.942}$$

$$g_3 = D_3 q_0 + D_2 q_1 = d_3 q_0 z^3 + d_2 q_1 z^4 \tag{6.943}$$

$$g_4 = D_3 q_1 = d_3 q_1 z^3 \tag{6.944}$$

Hence, we have

$$R_1 = \left(\frac{Q}{D}\right)(f_1 E_2 + f_2 E_3 + f_3 E_4 + f_4 E_5)$$

$$= \left[\frac{1}{z^4 (6J_2 d)}\right][a_3 z^3 + a_2 z^2 + a_1 z] \tag{6.945}$$

For convenience of operations, let

$$Y = \left(\frac{1}{z^3}\right)[a_3 z^2 + a_2 z + a_1] \tag{6.946}$$

and let $1/z = z$. Then we have

$$Y = a_1 x^3 + a_2 x^2 + a_3 x \tag{6.947}$$

where a_1, a_2, and a_3 may be positive, negative, or zero depending on q_0, q_1, and E_j. But is this so?

Consider a_3. It can be shown to be positive in at least two ways. Let us play around a little bit in the following manner first. Suppose a_3 is zero. Then

$$q_0 E_2 + q_1 E_3 = 0 \tag{6.948}$$

which leads to

$$\frac{q_0}{q_1} = \frac{-E_3}{E_2} \tag{6.949}$$

However, $q_0/q_1 = -a$; thus, from (6.949) we have

$$\frac{E_3}{E_2} = a \tag{6.950}$$

Invoking the definitions of E_2 and E_3 in (6.950), we have

$$a = \frac{2(e^2 + ea + a^2)}{3(e + a)} \tag{6.951}$$

which yields $ae + a^2 > 2e^2$. Since by definition $e = a + b$ and $a > 0$, $b > 0$, we have $e > 0$. The only possible root of (6.951) as a quadratic equation in e is $e = a$, which corresponds to the trivial load case where the loaded length $b = 0$ (i.e., zero load on the beam).

Suppose $a_3 < 0$. Then

$$q_0 E_2 + q_1 E_3 < 0 \tag{6.952}$$

which means $a > E_3/E_2$, or equivalently

$$ae + a^2 > 2e^2 \tag{6.953}$$

However, this is impossible, because by definition $a < e$ or $a = e$ at most. Therefore, unless $a = e$ is the case and thus $a_3 = 0$, we have $a_3 > 0$ in general.

Another proof is via the properties of E_j ($j = 2, 3, 4, 5$) established earlier. One of these properties is $E_{j+1} - aE_j > 0$ in general except when $e = a$, in which case $E_{j+1} - aE_j = 0$.

Next, we claim that $a_2 < 0$ except in the trivial case when $e = a$ with zero load.

Proof

It is given that

$$a_2 = \frac{-1}{2J_3}(q_0 E_3 + q_1 E_4) \tag{6.954}$$

which is equal to

$$a_2 = \frac{-1}{2J_3}(q_1)(-aE_3 + E_4) \tag{6.955}$$

because $q_0/q_1 = -a$.

Since $(-aE_3 + E_4) > 0$ and $q_1 > 0$, we have from these two facts and (6.955) the result $a_2 < 0$.

Similarly, we can show that $a_1 > 0$ by using the established E_j property $E_{j+1} - aE_j > 0$ for all general cases excluding the trivial case $e = a$ (with zero load on the beam).

With the signs of all the a_j established, we are ready to tackle

$$Y' = a_3 + 2a_2 x + 3a_1 x^2 \tag{6.956}$$

The roots of the equation $Y' = 0$ are

$$x_1 = \frac{-a_2 + S^{1/2}}{3a_1} \tag{6.957}$$

$$x_2 = \frac{-a_2 - S^{1/2}}{3a_1} \tag{6.958}$$

and they are real and distinctive if

$$S = a_2^2 - 3a_1 a_3 > 0 \tag{6.959}$$

which means

$$3(q_0^2 E_3^2 + q_1^2 E_4^2) + 4q_0 q_1 E_3 E_4 > 2(q_0^2 E_2 E_4 + q_1^2 E_3 E_5 + q_0 q_1 E_2 E_5) \tag{6.960}$$

Regrouping the terms in (6.960) and dividing each term by the positive quantity q_1^2, we have

$$a^2(3E_3^2 - 2E_2 E_4) - a(4E_3 E_4 - 2E_2 E_5) + (3E_4^2 - 2E_3 E_5) > 0 \tag{6.961}$$

We have in (6.961) basically three terms and will concentrate on them one at a time as follows. For the first term, consider $A = (3E_3^2 - 2E_2 E_4)$. Substituting the definitions of the appropriate E_j, we obtain

$$A = \left[\frac{(e-a)^2}{12}\right][e^4 + 2e^3 a + 6e^2 a^2 + 2ea^3 + a^4] \tag{6.962}$$

$$= \frac{b^2 K}{12} > 0, \qquad\qquad e - a = b$$

Next, we consider $B = 4E_3E_4 - 2E_2E_5$ and obtain

$$B = \left(\frac{1}{15}\right) [2(e^4 + a^4 + e^3a + ea^3) + 7e^2a^2] > 0 \qquad (6.963)$$

Finally, we look at $C = 3E_4^2 - 2E_3E_5$ and find

$$C = \left(\frac{13}{240}\right)(e^8 + a^8) - e^3a^3\left[\left(\frac{3}{8}\right)ea - \left(\frac{2}{15}\right)(e^2 + a^2)\right] \qquad (6.964)$$

which is pretty messy. However, we can make a nice estimate of C while proving $C > 0$. The trick is to note that $(e - a)^2 = e^2 + a^2 - 2ea > 0$, thereby obtaining the desired inequality. Thus,

$$
\begin{aligned}
2E_3E_5 &= \left(\frac{2}{15}\right)[e^8 + a^8 - e^3a^3(e^2 + a^2)] \\
&< \left(\frac{2}{15}\right)[e^8 + a^8 - 2e^4a^4] \\
&< \left(\frac{3}{16}\right)[e^8 + a^8 - 2e^4a^4] \\
&= 3E_4^2
\end{aligned}
\qquad (6.965)
$$

and a nice estimate of C is possible by noting $(16/15)E_4^2 > E_3E_5$.

It is interesting to note, in passing, that

$$E_3^2 < E_2E_4 \qquad (6.965A)$$

A sketch of the proof of (6.965) is as follows. We start off with

$$4e^2a^2 - 2ea(a^2 + e^2) < 4e^2a^2 - 2ea(2ea) = 0 < (e^2 - a^2)^2 \qquad (6.966)$$

to get

$$6e^2a^2 < e^4 + a^4 + 2ea(a^2 + e^2) \qquad (6.967)$$

which leads to

$$
\begin{aligned}
8(e^4 &+ a^4 + 2e^3a + 2ea^3 + 3e^2a^2) \\
&< 9[e^4 + a^4 + 2(e^2a^2 + e^3a + ea^3)]
\end{aligned}
\qquad (6.968)
$$

and eventually the desired result in (6.965).

This proof was purposefully done in a manner different from the usual approach to add some spice to the treatment in order to raise interest. There are, of course, other approaches to devising a proof.

Let us return to the main stream of development, namely (6.959) and (6.961), by saying that

$$S = a^2A - aB + C > 0 \quad \text{or} \quad S = 0 \tag{6.969}$$

is the necessary condition for $Y' = 0$ to have real roots. Note, however, that we did not say that Y' has real roots, nor did we say the contrary either. We note that A, B, and C as well as the dimension a are all positive.

This should wrap up the story of R_1 and let us move on to R_4 given by

$$R_4 = \frac{g_1E_2 + g_2E_3 + g_3E_4 + g_4E_5}{D} \tag{6.970}$$

where the numerator can be written as $b_1z^5 + b_2z^4 + b_3z^3$, and the denominator is

$$D = z^6d \tag{6.971}$$

where

$$d = 3B_2^2 + 4(B_1 + B_3)B_2 + 4B_1B_3 \tag{6.972}$$

with B_i $(i = 1, 2, 3)$ given by (6.931A).

Thus,

$$R_4 = \frac{b_1z^{-1} + b_2z^{-2} + b_3z^{-3}}{d} \tag{6.973}$$

where

$$b_1 = d_1(q_0E_2 + q_1E_3) < 0 \tag{6.974}$$

$$b_2 = d_2(q_0E_3 + q_1E_4) < 0 \tag{6.975}$$

$$b_3 = d_3(q_0E_4 + q_1E_5) > 0 \tag{6.976}$$

with the signs of the b's determined by those of the d's.

Now we introduce a new variable $x = 1/z$ and consider the function

$$Y = b_1x + b_2x^2 + b_3x^3 \tag{6.977}$$

We have

$$Y' = b_1 + 2b_2x + 3b_3x^2 \tag{6.978}$$

Thus, the roots of $Y' = 0$ are

$$x_1 = \frac{-b_2 + S^{1/2}}{3b_3} \tag{6.979}$$

$$x_2 = \frac{-b_2 - S^{1/2}}{3b_3} \tag{6.980}$$

where

$$S = b_2^2 - 3b_1b_3 \tag{6.981}$$

Since $b_1 b_3 < 0$, we have $S > 0$ and two real roots of $Y' = 0$. However, because $S^{1/2} > b_2$, we have only one positive root given by (6.979). Moreover, $Y'' = 2b_2 + 6b_3 x$ and $Y'' > 0$ at x_1, so Y has a minimum at x_1. From (6.973) and (6.977), we know that Y is proportional to R_4. Therefore, R_4 has a minimum at x_1 also.

6.17. A CONCENTRATED COUPLE AT AN ARBITRARY POINT ON AN EXTERIOR SPAN OF A THREE-SPAN CONTINUOUS BEAM: CONSTANT J

The reaction R_1 is

$$R_1 = \frac{HK - FZ}{D} \tag{6.982}$$

where

$$H = \frac{L_1 L_2 L_3}{6J} \tag{6.983}$$

$$K = -M_0 \frac{\left(\frac{2L_2 L_3}{J} + 3 \frac{L_3^2 - c^2}{J} \right)}{6} \tag{6.984}$$

$$F = \frac{-L_1 L_2 M_0}{6J} \tag{6.985}$$

$$Z = \frac{L_3^2 (L_2 + L_3)}{3J} \tag{6.986}$$

$$D = \frac{d}{J^2} \quad \text{with } d = \left(\frac{L_1 L_3}{6} \right)^2 [3L_2^2 + 4L_2(L_1 + L_3) + 4L_1 L_3] \tag{6.987}$$

Substituting the entities given by (6.983–6.987) into (6.982), we obtain

$$R_1 = \frac{M_0 L_2 (L_1 L_3) (3c^2 - L_3^2)}{T} \tag{6.988}$$

where

$$T = 3L_2^2 + 4L_2(L_1 + L_3) + 4L_1 L_3 \tag{6.989}$$

Note that, strictly and even pedantically speaking, given the sign of M_0, the entity $(3c^2 - L_3^2)$ determines the sign of R_1 since all the other factors in the expression for

R_1 are positive. $(3c^2 - L_3^2)$ is a critical factor because it dictates whether R_1 is zero as well. Thus, we will call this span length L_3, which renders R_1 zero as the critical span length. Thus, for $M_0 > 0$, we have the following situation: $R_1 < 0$, = 0, or > 0 when

$$3c^2 - L_3^2 < 0, = 0, \text{ or } > 0 \qquad (6.990)$$

respectively.

We can look at (6.990) from at least two aspects. One is to hold L_3 fixed and see which value or values of c will satisfy the first or last of the three relations in (6.990), with the second relation in (6.990) defining the critical span length. The other aspect is to hold c fixed and see which span lengths L_3 fit the category where $R_1 < 0$ and which ones fit the category where $R_1 > 0$. We could, of course, do an entirely analogous analysis and correspondingly draw a set of conclusions similar to the above.

Additionally, it is interesting to note that the absolute value of R_1 decreases as the value of L_3 increases through the contributing factors.

Let us move on to R_4, which is displayed in (6.991) as

$$R_4 = \frac{FN - GK}{D} \qquad (6.991)$$

where

$$N = H = \frac{L_1 L_2 L_3}{6J} \qquad (6.992)$$

$$F = \frac{-L_1 L_2 M_0}{6J} \qquad (6.993)$$

$$G = \frac{L_1^2 (L_1 + L_2)}{3J} \qquad (6.994)$$

Substituting (6.992–6.994) into (6.991) and simplifying, we obtain

$$R_4 = \frac{M_0 S}{L_3^2 T} \qquad (6.995)$$

where

$$S = 4L_1 L_2 L_3 + 6(L_1 + L_2)(L_3^2 - c^2) + 3L_2^2 L_3 \qquad (6.996)$$

and is given in Formula (6.989).

Note that R_4, like R_1, is again independent of J. However, unlike R_1, the reaction R_4 is nonzero unless $M_0 = 0$. Moreover, $R_4 < 0$ or $R_4 > 0$ depends on $M_0 < 0$ or $M_0 > 0$ respectively, since S and T are both strictly positive. From (6.995) and (6.996), we see that as c increases, R_4 decreases. When $c = 0$,

$$S = 4L_1L_2L_3 + 6(L_1 + L_2)L_3^2 + 3L_2^2L_3 \tag{6.997}$$

and this is a maximum S and R_4 when c is viewed as the variable. When $c = L_3$, the entity S is simplified to

$$S = 4L_1L_2L_3 + 3L_2^2L_3 \tag{6.998}$$

which corresponds to a minimum S and R_4 when c is the variable.

6.18. A CONCENTRATED COUPLE AT AN ARBITRARY POINT ON AN EXTERIOR SPAN OF A THREE-SPAN CONTINUOUS BEAM: CONSTANT SPAN LENGTH

Here, $R_1 = (HK - FZ)/D$ and $R_4 = (FN - GK)/D$ are given as in the last section formally. However, the entities have different characteristics. Hence,

$$D = L^6 d, \quad \text{with } d = \left(\frac{1}{36}\right)\left[\frac{3}{J_2^2} + 4\frac{\left(\frac{1}{J_1} + \frac{1}{J_3}\right)}{J_2} + \frac{4}{J_1 J_3}\right] \tag{6.999}$$

$$F = \frac{-L^2 M_0}{6J_2} \tag{6.1000}$$

$$G = L^3 \frac{\left(\frac{1}{J_1} + \frac{1}{J_2}\right)}{3} \tag{6.1001}$$

$$H = \frac{L^3}{6J_2} = N \tag{6.1002}$$

$$K = \frac{-M_0}{6J_2}(5L^2 - 3c^2) \tag{6.1003}$$

$$Z = L^3 \frac{\left(\frac{1}{J_2} + \frac{1}{J_3}\right)}{3} \tag{6.1004}$$

and

$$R_1 = \frac{\left(\dfrac{a_1}{L} + \dfrac{a_2}{L^3}\right)}{d} \tag{6.1005}$$

where

$$a_1 = M_0 \frac{\left(\dfrac{2}{J_2} - \dfrac{3}{J_3}\right)}{36J_2} \tag{6.1006}$$

$$a_2 = \frac{M_0 c^2}{12 J_2 J_3} > 0 \tag{6.1007}$$

Let $x = 1/L$ and $R_1 = Y/d$ and consider Y as a function of x:

$$Y = a_1 x + a_2 x^3 \tag{6.1008}$$

Then

$$Y' = a_1 + 3a_2 x^2 \tag{6.1009}$$

Examining a_1, we find that $a_1 < 0, = 0$, or > 0 when

$$2J_3 - 3J_2 < 0, = 0, \text{ or } > 0 \tag{6.1010}$$

respectively.

Note that $a_2 > 0$, so $Y'' = 6a_2 x > 0$. Let us look at $Y' = 0$. This happens when

$$x^2 = \frac{-a_1}{3a_2} \tag{6.1011}$$

which must be positive because $x^2 > 0$ for any real x_1, not to mention $x = 1/L$ and $L > 0$. Hence, $a_1 < 0$ is required and from (16.1010) we have

$$2J_3 < 3J_2 \tag{6.1012}$$

as the condition to be met. Here, $x = (-a_1/3a_2)^{1/2}$ and Y has a minimum.

For a_1 which does not satisfy (6.1012), we have $Y' > 0$. This is because of the following. When $a_1 = 0$, the condition $Y' = 0$ implies that $x = 0$, which is a trivial case, meaning $Y' = 3a_2 x > 0$ in general. Now when $a_1 > 0$, both terms of Y' are positive. As a result, $Y' > 0$. In short, Y is an increasing function of x when (6.1012) is not satisfied.

Next, we look at R_4. Using a similar approach as we did for R_1, we have

$$R_4 = \frac{X}{d} \tag{6.1013}$$

where

$$X = a_1 x + a_2 x^3 \tag{6.1014}$$

with

$$x = \frac{1}{L}, \quad a_1 = \left(\frac{M_0}{36}\right)\left[10\left(\frac{\dfrac{1}{J_1} + \dfrac{1}{J_2}}{J_3}\right) - \frac{1}{J_2^2}\right], \tag{6.1015}$$

$$a_2 = -M_0 c^2 \left[\frac{\dfrac{1}{J_1} + \dfrac{1}{J_2}}{\dfrac{6}{J_3}}\right]$$

Note that $a_1 < 0, = 0,$ or > 0 when

$$\left(\frac{10}{J_3}\right)\left(\frac{1}{J_1} + \frac{1}{J_2}\right) - \frac{1}{J_2^2} < 0, = 0, \text{ or } > 0 \tag{6.1016}$$

respectively.

As for a_2, we have $a_2 < 0$ for $M_0 > 0$ and $a_2 > 0$ for $M_0 < 0$.

Let us look at $X' = a_1 + 3a_2 x^2 = 0$ in the following manner.

Case 1. $a_1 > 0$

$X' = 0$ is possible at $x^2 = -a_1/(3a_2) > 0$ because $a_2 < 0$ when $M_0 > 0$. $X'' = 6a_2 x < 0$. Hence, at $x = [-a_1/(3a_2)]^{1/2}$, X has a maximum, and so does R_4.

Case 2. $a_1 = 0$

Here, $X = a_2 x^3$, $X' = 3a_2 x^2$, and $X' = 0$ when $x = 0$, which is meaningless. Both X and X' have the same sign as that of a_2. Thus, when $M_0 > 0$, we have X as a decreasing function of x and therefore an increasing function of L, and of course, when $M_0 < 0$, the opposite is true.

Case 3. $a_1 < 0$

In this case, $X' = a_1 + 3a_2 x^2 < 0$ if $M_0 > 0$ and X is a decreasing function of x. If, however, $M_0 < 0$, then $a_2 > 0$ and X' may be positive, negative, or zero. If $X' = 0$, then $x^2 = -a_1/(3a_2)$ and at this point $X'' = 6a_2 x > 0$, yielding a minimum for X_1 and for R_4 also.

6.19. A CONCENTRATED FORCE AT AN ARBITRARY POINT ON THE INTERIOR SPAN OF A THREE-SPAN CONTINUOUS BEAM: CONSTANT J

The reaction R_1 is given by

$$R_1 = VC(A_0 + A_1 s)P \tag{6.1017}$$

where

$$V = s\left[\frac{L_2 - s}{6J_2 L_2}\right] \tag{6.1018}$$

$$C = L_1 L_3^2 (6D) \tag{6.1019}$$

$$A_0 = \frac{-2L_2 L_3}{J} \tag{6.1020}$$

$$A_1 = -\frac{3L_2 + L_3}{J} \tag{6.1021}$$

$$D = \frac{d}{J^2} \tag{6.1022}$$

with

$$d = \left[\frac{L_1 L_3}{6}\right]^2 [3L_2^2 + 4(L_1 + L_3)L_2 + 4L_1 L_3] \tag{6.1023}$$

For constant P, consider

$$\frac{R_1}{P} = FBL_3 \tag{6.1024}$$

where

$$F = s(L_2 - s)\frac{L_1 L_3^2}{36L_2 d} \tag{6.1025}$$

$$B = -2L_2 L_3 - s(3L_2 + 2L_3) \tag{6.1026}$$

are all independent of J. The ratio R_1/P is, therefore, also free of J.

Now let us look at R_4/P, given by

$$\frac{R_4}{P} = VC'(D_0 + D_1 s) \tag{6.1027}$$

where

$$C' = C\left(\frac{L_1}{L_3}\right) = \frac{L_1^2 L_3}{6D} \tag{6.1028}$$

$$D_0 = -L_2\left(\frac{4L_1 + 3L_2}{J}\right) \tag{6.1029}$$

$$D_1 = \frac{2L_1 + 3L_2}{J} \tag{6.1030}$$

Thus,

$$\frac{R_4}{P} = FGL_1 \tag{6.1031}$$

where

$$G = -L_2(4L_1 + 3L_2) + s(2L_1 + 3L_2) \tag{6.1032}$$

is independent of J.

Note that F may be zero or positive. $F = 0$ when $s = L_2$, in which case both R_1 and R_2 are zero. Note also that $B < 0$. Thus, $R_1 < 0$ when F is nonzero.

Can we have $G = 0$? Suppose we do. $G = 0$ when

$$s = L_2\left(\frac{4L_1 + 3L_2}{2L_1 + 3L_2}\right) > L_2 \tag{6.1033}$$

which has no physical meaning. Therefore, the only possibilities are $G < 0$ or $G > 0$. However, the latter is not possible in view of (6.1033), so $G < 0$. This means that $R_4 < 0$, except that it is zero when $F = 0$, in which case $R_1 = 0$ also.

To summarize, we know that both R_1 and R_4 are in general negative except when $F = 0$, in which case they are both zero at $s = L_2$.

6.20. A CONCENTRATED FORCE AT AN ARBITRARY POINT ON THE INTERIOR SPAN OF A THREE-SPAN CONTINUOUS BEAM: CONSTANT SPAN LENGTH

Again, we are considering R_1/P in the form presented in (6.1017) in the last section. However, the entities involved in R_1/P have different elements. Thus,

$$V = \frac{s(L - s)}{6J_2 L} \tag{6.1034}$$

$$C = \frac{L^3}{6D} \tag{6.1035}$$

$$A_0 = \frac{-2L^2}{J_3} \tag{6.1036}$$

$$A_1 = L\left(\frac{3}{J_2} + \frac{2}{J_3}\right) \tag{6.1037}$$

Thus,

$$\frac{R_1}{P} = a_1 L^{-2} + a_2 L^{-3} \tag{6.1038}$$

where

$$a_1 = \frac{-2K}{J_3} \tag{6.1039}$$

$$a_2 = -sK\left(\frac{3}{J_2} + \frac{2}{J_3}\right) \tag{6.1040}$$

with

$$K = \frac{s(L - s)}{36J_2 d} \tag{6.1041}$$

which is nonnegative. Note that $K = 0$ only when $s = L$ or $s = 0$, with the result $a_1 = 0$, $a_2 = 0$, and $R_1 = 0$.

Let $R_1/P = Y$ and $1/L = x$ and consider Y as a function of x. Thus,

$$Y = a_1 x^2 + a_2 x^3 \tag{6.1042}$$

and

$$Y' = 2a_1 x + 3a_2 x^3 \tag{6.1043}$$

$Y' = 0$ implies $x = 0$ or $2a_1 + 3a_2 x = 0$. The latter means

$$x = \frac{-2a_1}{3a_2} \tag{6.1044}$$

which is not admissible, because the ratio $2a_1/(3a_2)$ in (6.1044) is positive and a minus sign in front of it renders x negative. Note that the case where $s = 0$ renders $K = 0$, and hence $a_1 = 0$, $a_2 = 0$, as well as $R_1 = 0$, is, therefore, excluded from consideration when dealing with (6.1044) with regard to the admissibility of x given in (6.1044).

Another way to look at $Y' = 0$ is to focus on $a_1 = 0$ and $a_2 = 0$, which yield $Y' = 0$ regardless of the value of x. However, $a_1 = 0$ and $a_2 = 0$ at the same time means that $K = 0$ and thus $s = 0$, $Y = 0$, and $R_1 = 0$ also.

In short, $a_1 < 0$ and $a_2 < 0$ except when $K = 0$. It is clear from the expression for Y' that $Y' < 0$ for all x, and the same statement can be made about Y and thus

R_1 for that matter. Even $Y'' < 0$ holds for nonzero a_1 and a_2, and once again, when $a_1 = 0$ and $a_2 = 0$ at the same time, we have $Y'' = 0$ also, along with Y', Y, and R_1.

Let us do

$$\frac{R_4}{P} = VC'(D_0 + D_1 s) = b_1 L^{-2} + b_2 L^{-3} \tag{6.1045}$$

where

$$b_1 = -K\left(\frac{1}{J_1} + \frac{3}{J_2}\right) < 0 \tag{6.1046}$$

$$b_2 = Ks\left(\frac{2}{J_2} + \frac{3}{J_3}\right) > 0 \tag{6.1047}$$

in general except when $K = 0$, in which case they both are zero and so is R_4. Thus, for nonzero K, we have, by letting $1/L = x$ and $X = b_1 x^2 + b_2 x_1^3$, the result

$$X' = 2b_1 x + 3b_2 x^2 \tag{6.1048}$$

We see that $X' = 0$ when $x = 0$ or

$$x = \frac{-2b_1}{3b_2} \tag{6.1049}$$

which is possible to be positive because the entities b_1 and b_2 are of opposite signs for nonzero K.

We have $X'' = 2b_1 + 6b_2 x$, so $X'' > 0$ at the value of x given by (6.1049). Therefore, X has a minimum there, and everything else regarding an increase or a decrease of function X falls into place. When will R_4 be positive, zero, or negative? Well, $X = 0$ implies $R_4 = 0$ and $b_1 + b_2 x = 0$, and the latter means either $b_1 = b_2 = 0$ when $K = 0$ or

$$x = \frac{-b_1}{b_2} \tag{6.1050}$$

for nonzero K. Note that (6.1050) is admissible because here b_1 and b_2 are of opposite signs.

Naturally, $X > 0$ when $x > -b_1/b_2$ and $X < 0$ when $x < -b_1/b_2$.

6.21. UNIFORM LOAD ON THE INTERIOR SPAN OF A THREE-SPAN CONTINUOUS BEAM: CONSTANT J

The reaction R_1 is given by

$$R_1 = wSL_3 Z_1 \tag{6.1051}$$

where w is the uniform load and

$$S = \frac{L_1 L_3}{36 J L_2 D} \qquad (6.1052)$$

$$Z_1 = (A_0 L_2 + A_1 L_2^2) E_2 - (A_0 + 2 A_1 L_2) E_3 + A_1 E_4 = \frac{-X_1}{J} \qquad (6.1053)$$

with

$$A_0 = \frac{-2 L_2 L_3}{J} \qquad (6.1054)$$

$$A_1 = \frac{-(3 L_2 + 2 L_3)}{J} \qquad (6.1055)$$

$$D = \frac{d}{J^2} \qquad (6.1056)$$

$$X_1 = L_2^2 (3 L_2 + 4 L_3) E_2 - 6 L_2 (L_2 + L_3) E_3 + (3 L_2 + 2 L_3) E_4 \qquad (6.1057)$$

and d is as defined in Section 6.13 on the same structure regardless of load type.

As a result of the above, we have

$$R_1 = \frac{w L_1 L_3^2 X_1}{36 L_2 d} \qquad (6.1058)$$

Note that X_1 is independent of J and so is R_1. Moreover, X_1 has an upper bound X_{1U} and a lower bound X_{1L} as:

$$X_{1U} = (L_3 + L_2)[4 L_2^2 E_2 - 6 L_2 E_3 + 3 E_4] \qquad (6.1059)$$

$$X_{1L} = (L_2 + L_3)[3 L_2^2 E_2 - 6 L_2 E_3 + 2 E_4] \qquad (6.1060)$$

From the expression for X_1, it is conceivable that $X_1 = 0$, < 0, or > 0 for certain choices of L_2 and L_3 for a given entity a, and we can treat X_1 as a function of L_2 or L_3 only and get results.

Note that R_1 may be positive, negative, or zero through X_1; also, R_1 may have its upper bound or lower bound through those for X_1.

Now we can do something similar to R_4.

$$R_4 = w S L_1 Z_4 = \frac{w L_1^2 L_3 X_4}{36 L_2 d} \qquad (6.1061)$$

where

$$X_4 = -2 L_1 L_2^2 E_2 - 3 L_2^2 E_3 + (2 L_1 + 3 L_2) E_4 \qquad (6.1062)$$

is independent of J and so is R_4. Again, it is conceivable that $X_4 = 0$, < 0, or > 0 and the same remarks apply to R_4.

Similar to what we did for X_1 above, we can obtain an upper bound X_{4U} and a lower bound X_{4L} for X_4 as follows:

$$X_{4U} = -L_2^2(2L_1E_2 + 3E_3) + 3(L_1 + L_2)E_4 \qquad (6.1063)$$

$$X_{4L} = -3L_2^2(L_1E_2 + E_3) + 2(L_1 + L_2)E_4 \qquad (6.1064)$$

Naturally, R_4 will have its upper and lower bounds through those for X_4. Again, we can treat X_4 as a function of either L_1 or L_2 only and obtain useful and interesting results. Similar treatment of X_{4U} or X_{4L} can be done with ease also.

Note that we can obtain other upper and lower bounds for X_4 and X_1 for that matter as the need warrants.

6.22. UNIFORM LOAD ON THE INTERIOR SPAN OF A THREE-SPAN CONTINUOUS BEAM: CONSTANT SPAN LENGTH

The reaction R_1 is given as

$$R_1 = wSL_3Z_1 \qquad (6.1065)$$

where

$$S = \frac{L}{36J_2D} \qquad (6.1066)$$

$$Z_1 = -\left(\frac{3}{J_2} + \frac{4}{J_3}\right)L^3E_2 + 6\left(\frac{1}{J_2} + \frac{1}{J_3}\right)L^2E_3 - \left(\frac{3}{J_2} + \frac{2}{J_3}\right)LE_4 \qquad (6.1067)$$

Thus,

$$R_1 = \frac{wY}{36J_2d} \qquad (6.1068)$$

where

$$Y = a_1x + a_2x^2 + a_3x^3, \qquad x = \frac{1}{L} \qquad (6.1069)$$

with

$$a_1 = -E_2\left(\frac{3}{J_2} + \frac{4}{J_3}\right) \qquad (6.1070)$$

$$a_2 = 6E_3\left(\frac{1}{J_2} + \frac{1}{J_3}\right) \qquad (6.1071)$$

$$a_3 = E_4 \left(\frac{3}{J_2} + \frac{2}{J_3} \right) \tag{6.1072}$$

Note that in general $a_1 < 0$, $a_2 > 0$, and $a_3 < 0$, and they are zero only when $E_j = 0$, which means that $e = a$, signifying a trivial case with zero load. It is interesting to see that Y has an upper bound Y_U and a lower bound Y_L:

$$Y_U = x \left(\frac{1}{J_2} + \frac{1}{J_3} \right) [-3E_2 + 6xE_3 - 2x^2 E_4] \tag{6.1069A}$$

$$Y_L = x \left(\frac{1}{J_2} + \frac{1}{J_3} \right) [-4E_2 + 6xE_3 - 3x^2 E_4] \tag{6.1069B}$$

Consider $Y' = a_1 + 2a_2 x + 3a_3 x^2$ and $Y'' = 2a_2 + 6a_3 x$. We see that $Y' = 0$ when

$$x = \frac{-a_2 + S^{1/2}}{3a_3} \tag{6.1073}$$

or when

$$x = \frac{-a_2 - S^{1/2}}{3a_3} \tag{6.1074}$$

where

$$S = a_2^2 - 3a_1 a_3 \tag{6.1075}$$

For the real roots of $Y' = 0$, we need

$$a_2^2 - 3a_1 a_3 \geq 0 \tag{6.1076}$$

We have two positive roots given by (6.1073) and (6.1074). For the first one, we have $Y'' > 0$, so Y has a minimum there. For the second one, we have $Y'' < 0$, so Y has a maximum there.

Note that $Y = 0$ at $x = 0$, or $a_1 + a_2 x + a_3 x^2 = 0$. The former means that L is infinitely long, which is not admissible. The latter is satisfied when

$$x = \frac{-a_2 + T^{1/2}}{2a_3} \tag{6.1077}$$

or

$$x = \frac{-a_2 - T^{1/2}}{2a_3} \tag{6.1078}$$

where

$$T = a_2^2 - 4a_1 a_3 \tag{6.1079}$$

must be nonnegative for real roots. The question is: Can $T > 0$ or $= 0$? It is worth-while to explore.

We now proceed to work on R_4, which is expressed as

$$R_4 = wSLZ_4 = \frac{w}{(36\,J_2\,d)\,X} \tag{6.1080}$$

where

$$X = b_1 x + b_2 x^2 + b_3 x^3, \qquad x = \frac{1}{L} \tag{6.1081}$$

with

$$b_1 = \frac{-2E_2}{J_1} < 0 \tag{6.1082}$$

$$b_2 = \frac{-3E_3}{J_2} < 0 \tag{6.1083}$$

$$b_3 = E_4\left(\frac{2}{J_2} + \frac{3}{J_2}\right) > 0 \tag{6.1084}$$

Now

$$X' = b_1 + 2b_2 x + 3b_3 x^2 \tag{6.1085}$$

$$X'' = 2b_2 + 6b_3 x \tag{6.1086}$$

so $X' = 0$ when

$$x_1 = \frac{-b_2 + U^{1/2}}{3b_3} \tag{6.1087}$$

or

$$x_2 = \frac{-b_2 - U^{1/2}}{3b_3} \tag{6.1088}$$

where

$$U = b_2^2 - 3b_1 b_3 \tag{6.1089}$$

is positive because $b_1 b_3 < 0$ and ensures that we have two real roots for $X' = 0$. However, only the one root given by (6.1087) is positive. At this x, we have $X'' > 0$; therefore, X has a minimum here.

Let us look at $X = 0$. This happens when $x = 0$, which is a trivial case, or when

$$x_1 = \frac{-b_2 + V^{1/2}}{2b_3} \tag{6.1090}$$

or

$$x_2 = \frac{-b_2 - V^{1/2}}{2b_3} \tag{6.1091}$$

is satisfied where

$$V = b_2^2 - 4b_1 b_3 \tag{6.1092}$$

is positive, as required for real roots. Again, only one root given by (6.1090) is positive and therefore admissible. Also, when $x > x_1$ given in (6.1090), we have $X > 0$ and $R_4 > 0$. Naturally, when $x < x_1$, we have $X < 0$ and $R_4 < 0$.

6.23. TRIANGULAR LOAD ON THE INTERIOR SPAN OF A THREE-SPAN CONTINUOUS BEAM: CONSTANT J

Here we have

$$A_0 = \frac{-2L_2 L_3}{J} = \frac{a_0}{J} < 0 \tag{6.1093}$$

$$A_1 = \frac{-(3L_2 + 2L_3)}{J} = \frac{a_1}{J} < 0 \tag{6.1094}$$

$$D_0 = \frac{-L_2(4L_1 + 3L_2)}{J} = \frac{d_0}{J} < 0 \tag{6.1095}$$

$$D_1 = \frac{2L_1 + 3L_2}{J} = \frac{d_1}{J} > 0 \tag{6.1096}$$

$$U = \frac{w}{6bJ} = \frac{u}{J} > 0 \tag{6.1097}$$

$$D = \frac{d}{J^2} > 0 \tag{6.1098}$$

Thus,

$$Z_1 = -aU = \frac{z_1}{J} < 0 \tag{6.1099}$$

$$Z_2 = \left(1 + \frac{a}{L_2}\right)U = \frac{z_2}{J} > 0 \tag{6.1100}$$

$$Z_3 = \frac{-U}{L_2} = \frac{z_3}{J} < 0 \tag{6.1101}$$

where

$$z_1 = \frac{-aw}{6b} < 0 \tag{6.1102}$$

$$z_2 = \left[1 + \frac{a}{L_2}\right]\frac{w}{6b} > 0 \tag{6.1103}$$

$$z_3 = \frac{w}{6bL_2} < 0 \tag{6.1104}$$

Also,

$$C = \frac{L_1 L_3^2 J^2}{6d} = kJ^2 > 0 \tag{6.1105}$$

$$C' = k'J^2 > 0 \tag{6.1106}$$

where

$$k = \frac{L_1 L_3^2 J}{6d} > 0 \tag{6.1107}$$

$$k' = k\left(\frac{L_1}{L_3}\right) > 0 \tag{6.1108}$$

Thus,

$$R_1 = k[b_1 E_2 + b_2 E_3 + b_3 E_4 + b_4 E_5] \tag{6.1109}$$

where

$$b_1 = a_0 z_1 > 0 \tag{6.1110}$$

$$b_2 = a_0 z_2 + a_1 z_1 \tag{6.1111}$$

$$b_3 = a_0 z_3 + a_1 z_2 \tag{6.1112}$$

$$b_4 = a_1 z_3 > 0 \tag{6.1113}$$

Similarly,

$$R_4 = k'[f_1 E_2 + f_2 E_3 + f_3 E_4 + f_4 E_5] \tag{6.1114}$$

where

$$f_1 = d_0 z_1 > 0 \tag{6.1115}$$

$$f_2 = d_0 z_2 + d_1 z_1 < 0 \tag{6.1116}$$

$$f_3 = d_0 z_3 + d_1 z_2 > 0 \tag{6.1117}$$

$$f_4 = d_1 z_3 < 0 \tag{6.1118}$$

Note that k, k', and all the b's, f's, and E's are independent of J and so are R_1 and R_4.

6.24. TRIANGULAR LOAD ON THE INTERIOR SPAN OF A THREE-SPAN CONTINUOUS BEAM: CONSTANT SPAN LENGTH

The reaction R_1 is given by

$$R_1 = U\left(\frac{b_2x + b_1x^2 + b_0x^3}{6d}\right) \tag{6.1119}$$

where

$$x = \frac{1}{L} \tag{6.1120}$$

$$b_2 = \frac{2(aE_2 - E_3)}{J_3} < 0 \tag{6.1121}$$

$$b_1 = \frac{3(aE_3 - E_4)}{J_2} < 0 \tag{6.1122}$$

$$b_0 = -\left(\frac{3}{J_2} + \frac{2}{J_3}\right)(aE_4 - E_5) > 0 \tag{6.1123}$$

Let

$$Y = c_1x + c_2x^2 + c_3x^3 \tag{6.1124}$$

with

$$c_1 = b_2, \qquad c_2 = b_1, \qquad c_3 = b_0 \tag{6.1125}$$

Consider

$$Y' = c_1 + 2c_2x + 3c_3x^2 \tag{6.1126}$$

$Y' = 0$ when

$$x_1 = \frac{-c_2 + S^{1/2}}{3c_3} \tag{6.1127}$$

or

$$x_2 = \frac{-c_2 - S^{1/2}}{3c_3} \tag{6.1128}$$

where

$$S = c_2^2 - 3c, \ c_3 \tag{6.1129}$$

Note that both roots of $Y' = 0$ are positive real ones and are thus admissible.

Now consider $Y'' = 2c_2 + 6c_3x$. We have $Y'' > 0$ when (6.1127) is satisfied, corresponding to a minimum Y. $Y'' < 0$ when (6.1128) is satisfied, signifying a maximum for Y here.

Note that Y can be positive, negative, or zero, and so can R. We see that $Y = 0$ when $x = 0$, which is a trivial case without physical meaning or

$$c_1 + c_2x + c_3x^2 = 0 \tag{6.1130}$$

the roots of which are given by

$$x_1 = \frac{-c_2 + T^{1/2}}{2c_3} \tag{6.1131}$$

$$x_2 = \frac{-c_2 - T^{1/2}}{2c_3} \tag{6.1132}$$

where

$$T = c_2^2 - 4c_1c_3 \tag{6.1133}$$

However, only the root given by (6.1131) is admissible. Moreover, $Y < 0$ when $x < x_1$, and $Y > 0$ when $x > x_1$. The former corresponds to $R_1 < 0$, while the latter corresponds to $R_1 > 0$.

Now we move on to R_4, which is given, from Section 6.10, as

$$R_4 = C(f_2L^2 + f_1L + f_0) \tag{6.1134}$$

where

$$f_2 = d_0U(E_3 - aE_2) < 0 \tag{6.1135}$$

$$f_1 = (d_1 - d_0)U(E_4 - aE_3) > 0 \tag{6.1136}$$

$$f_0 = d_1U(aE_4 - E_5) < 0 \tag{6.1137}$$

with

$$d_0 = -\left(\frac{4}{J_1} + \frac{3}{J_2}\right) < 0, \qquad d_1 = \left(\frac{2}{J_1} + \frac{3}{J_2}\right) > 0 \tag{6.1138}$$

Let

$$\frac{1}{L} = x, \qquad Y = f_2x + f_1x^2 + f_0x^3 \tag{6.1139}$$

Then,

$$Y' = f_2 + 2f_1 x + 3f_0 x^2 \tag{6.1140}$$

and $Y' = 0$ when

$$x = \frac{-f_1 + S^{\frac{1}{2}}}{3f_0} \tag{6.1141}$$

$$x = \frac{-f_1 - S^{\frac{1}{2}}}{3f_0} \tag{6.1142}$$

where

$$S = f_1^2 - 3f_0 f_2 \tag{6.1143}$$

We need $S > 0$ or $S = 0$ for real roots of $Y' = 0$. Do we have it? It is something worthwhile to pursue and is very similar to two problems: (1) when the load is on an exterior span and (2) uniform load on an interior span. The interested reader is encouraged to tackle this.

Let us pause for a moment and turn to Y. To this end, we rewrite Y as

$$Y = f_0(x^3 + g_1 x^2 + g_0 x) = x f_0 V \tag{6.1144}$$

The roots of $V = 0$ are

$$x_1 = \frac{-g_1 + T^{\frac{1}{2}}}{2} \tag{6.1145}$$

$$x_2 = \frac{-g_1 - T^{\frac{1}{2}}}{2} \tag{6.1146}$$

where

$$T = g_1^2 - 4g_0 \tag{6.1147}$$

However, these two roots given by (6.1145) and (6.1146) are negative even if they are real roots. Hence, V is nonzero also except at $x = 0$, which has no physical meaning. In fact, Y is negative because $V > 0$ and $f_0 < 0$ with x nonzero (and positive too).

In order to promote the reader's interest and learning efficiency, we will change pace as follows. Sections 6.25 and 6.26 are designed as exercises for the reader to find out whether the proposed approaches will give the correct solutions.

6.25. TWO-SPAN CONTINUOUS BEAM WITH SYMMETRY

An alternative way to treat the problem of a two-span continuous beam with symmetry (see Figure 6.1) is to consider the corresponding problem of a beam with a fixed end and a simply supported end as shown in Figure 6.2. The loads and material and section properties as well as the other structural characteristics are such that the

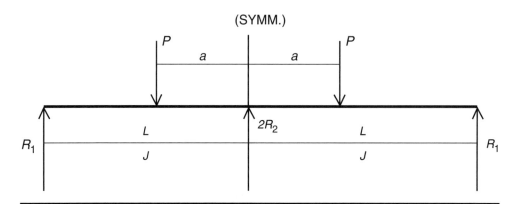

FIGURE 6.1. Two concentrated forces on a two-span continuous beam with symmetry

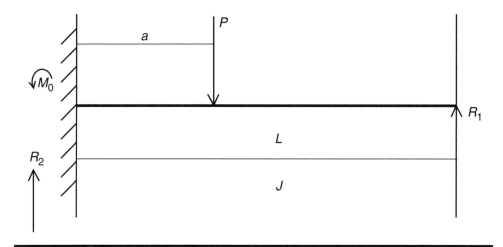

FIGURE 6.2. A concentrated force on a single-span beam with a fixed end and a simple supported end

problem of the structure represented by Figure 6.2 and its mirror image with the fixed end as the line of symmetry is equivalent to the problem in Figure 6.1.

Looking at Figure 6.1, the solution can be found from the discussion in Appendix F. Thus, all we have to do is first to identify which problem in Appendix F corresponds to a given two-span continuous beam problem with symmetry and then write down the answer. Since many different kinds of loads can be handled with Appendix F, just as many two-span continuous beams with symmetry can be taken care of.

Because of the nature of the problem of a two-span continuous beam with symmetry, the material and section properties as well as span lengths are constants for the whole structure. This is different from the corresponding problem for the case of a four-span continuous beam with symmetry, as will be seen next.

6.26. FOUR-SPAN CONTINUOUS BEAM WITH SYMMETRY

Basically, the problem of a four-span continuous beam with symmetry can be solved in a manner similar to a two-span continuous beam with symmetry in principle. Thus, we look at Figures 6.3 and 6.4 and observe immediately that the problem depicted in Figure 6.3 can be dealt with by considering Figure 6.4, which represents a second-degree statically indeterminate structural problem. Thus, for the two-span beam shown in Figure 6.4, we can take the two reactions at the two hinged supports as the unknowns for the system of equations resulting from the method of least work. As far as the material and section properties and the span lengths for this two-span structure are concerned, there are no restrictions. As a result of this generality, it is not necessary for the four-span structure depicted in Figure 6.3 to have constant

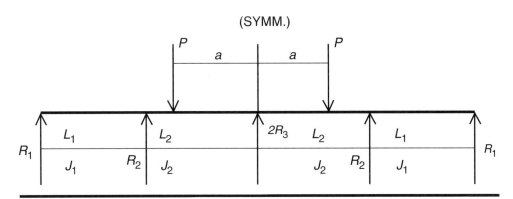

FIGURE 6.3. Two concentrated forces on a four-span continuous beam with symmetry

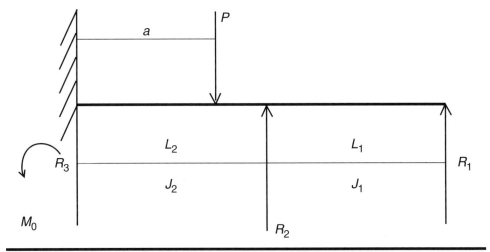

FIGURE 6.4. A concentrated force on a two-span beam with an exterior end fixed and the other supports hinged

material and section properties or constant span lengths. As long as the symmetry condition is satisfied, we have the right problem with the right solution. Thus, the categories regarding the material and section properties as well as span lengths mentioned before still apply here, and we have numerous interesting special cases in the usual way.

6.27. PRINCIPLE OF SUPERPOSITION

The principle of superposition is a fantastic way to generate or obtain solutions to new problems through solutions to old problems. Figures 6.5–6.7 are several interesting examples that show the application of the principle of superposition.

6.28. POSSIBLE DIRECTION FOR FUTURE WORK

Obviously, several important topics or problems are not included in this book. Some examples are (1) the various load cases for single-span beams with fixed ends and (2) a concentrated couple (a) on a single-span beam with a fixed end and a simply supported end and (b) on the interior span of a three-span continuous beam.

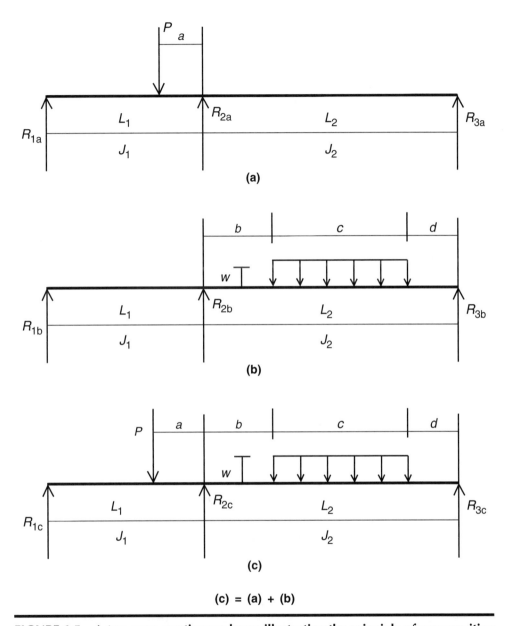

FIGURE 6.5. A two-span continuous beam illustrating the principle of superposition

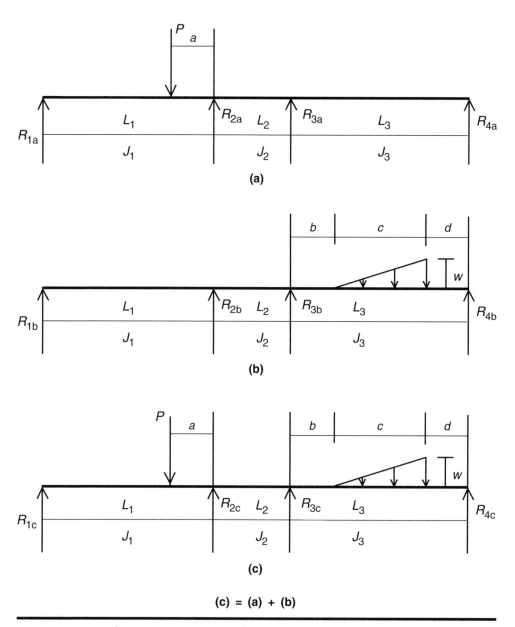

FIGURE 6.6. A three-span continuous beam illustrating the principle of superposition: Example 1

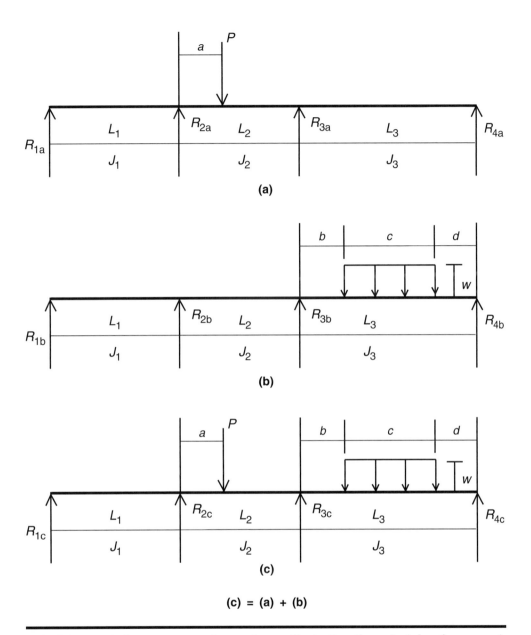

FIGURE 6.7. A three-span continuous beam illustrating the principle of superposition: Example 2

APPENDIX A.
SOME PROPERTIES OF $E_j - aE_k$

For the purpose of paving the way for future exploration, it is helpful to present some of the properties of several mathematical expressions which either are likely to occur (for example, for general j values) or we have already seen (for example, in cases where $j = 2, 3, 4, 5$) in functions involved in a parametric study or similar undertaking. Therefore, let us consider the following entities A_k of real numbers defined as

$$A_k = E_j - aE_k \qquad (A.1)$$

where

$$E_j = \frac{e^j - a^i}{j} \qquad (A.2)$$

with

$$k = j - 1 \qquad (A.3)$$

where j is a positive integer greater than 1, and

$$e = a + b \qquad (A.4)$$

where e is nonzero, and

$$a > 0 \quad \text{or} \quad a = 0 \qquad (A.5)$$

$$b > 0 \quad \text{or} \quad b = 0 \qquad (A.6)$$

Then, A_k has the properties listed below, in addition to the ones presented in Chapter 1 concerning $E_j - aE_k$.

First, A_k is proportional to several factors, namely b, C_k, e^k, and G_k, in a manner specified by Expressions (1.7–1.13), which were established utilizing different notations in Chapter 1 regarding $E_j - aE_k$. Thus,

$$A_k = C_k b e^k G_k \tag{A.7}$$

where

$$C_k = \frac{1}{jk} \tag{A.8}$$

$$G_k = k - (r + r^2 + \dots + r^k) \tag{A.9}$$

with

$$r = \frac{a}{e} \tag{A.10}$$

It is both interesting and important to note the following. From (A.4– A.6), we see that

$$0 < r < 1 \tag{A.11}$$

in general while

$$r = 1 \tag{A.12}$$

if and only if $a = e$ and $b = 0$.

Also, from (A.9), we have $G_k > 0$ when (A.11) is satisfied, whereas we obtain $G_k = 0$ when (A.12) is satisfied. In other words, $G_n > 0$ in general except when $r = 1$ and we have $G_n = 0$.

Second, A_k can be rewritten as

$$A_k = C_k(1 - r)e^j G_k \tag{A.13}$$

Third, G_k can be rewritten as

$$G_k = (1 - r)F_k \tag{A.14}$$

where

$$F_k = k + (k - 1)r + (k - 2)r^2 + \dots + r^{k-1} \tag{A.15}$$

Finally, G_k can further be rewritten as

$$G_k = k - S_k \tag{A.16}$$

with

$$S_k = r \frac{1 - r^k}{1 - r} \tag{A.17}$$

upon using Formula (A.18) for the first k term partial sum P_k of a geometric series in the form

$$c + cr + cr^2 + \ldots + cr^{k-1} \tag{A.18}$$

which is, with $c = r$ in our case, equal to

$$P_k = c\,\frac{1 - r^k}{1 - r} \tag{A.19}$$

Note that for $0 < r < 1$ (emphasizing that r is not equal to 1) when k is sufficiently large, then for all practical purposes we can approximate the exact value of P_k by

$$Q = \frac{c}{1 - r} \tag{A.20}$$

or we can treat Q as an upper bound for P_k with the property that the larger the value of k is, the closer the difference between P_k and Q becomes.

Next, we can start our exploration by keeping e fixed and considering A_k as a function of r for a given k.

We can also show that A_k is a decreasing function of r for $r \neq 1$. The proof is as follows.

Proof

Looking at (A.13) and paying special attention to the factors $(1 - r)$ and G_k in addition to the fact that both factors C_k and e^j are positive, we define

$$D_k = (1 - r)G_k \tag{A.21}$$

as a function of r and take the first derivative of D_k with respect to r. Thus,

$$D'_k = -j(1 - r) \tag{A.22}$$

We see from (A.22) that

$$D'_k < 0 \text{ for } 0 < r < 1 \tag{A.23}$$

which means that D_k is a decreasing function of r when $r \neq 1$. Additionally, we have

$$D'_k = 0 \text{ if } r = 1 \tag{A.24}$$

It is very interesting to note that if $r = 1$, then in addition to Expression (A.24) having something to do with zero, we also have $D_k = 0$ from Expression (A.21), $G_k = 0$ from as early as Expression (A.14), and even $A_k = 0$ from as early as Expression (A.13). We could trace this even further back if we choose to, but the point has been made and there is really no purpose in going any further back than Expression (A.13).

For application in the investigation of the behavior of beams under loads, the general case where $0 < r < 1$ is of interest for both practical usefulness and theoretical significance, whereas the special case where $r = 1$ corresponds to a trivial case of zero load on a beam.

Finally, let us revisit the special cases where $k = 2, 3, 4$ which we encountered frequently in Chapter 6. Thus,

$$A_2 = E_3 - aE_2 \tag{A.25}$$

$$A_3 = E_4 - aE_3 \tag{A.26}$$

$$A_4 = E_5 - aE_4 \tag{A.27}$$

are all members of a big family A_k, where k denotes the size of the family and can run from 1 all the way up to as high as one wants.

APPENDIX B.
SOME RELATIONS AMONG E_j ($j = 2, 3, 4, 5$)

Given the definitions of E_j as per Formula (A.1) in Appendix A, we have the following theorems.

THEOREM B.1

$$3E_4^2 > 2E_3E_5 \quad \text{for } e \neq a \tag{B.1}$$

Proof

From the definition of E_j, we have

$$E_4^2 = \frac{e^8 + a^8 - 2e^4a^4}{16} \tag{B.2}$$

$$E_3E_5 = \frac{(e^3 - a^3)(e^5 - a^5)}{15} = \frac{e^8 + a^8 - e^5a^3 - e^3a^5}{15}$$

$$= \frac{e^8 + a^8 - e^3a^3(e^2 + a^2)}{15} \tag{B.3}$$

However,

$$e^2 + e^2 > 2ae \tag{B.4}$$

so

$$e^3a^3(e^2 + a^2) > 2a^4e^4 \tag{B.5}$$

303

$$2E_3E_5 = 2\,\frac{e^8 + a^8 - e^3a^3(a^2 + e^2)}{15} < 2\,\frac{e^8 + a^8 - 2e^4a^4}{15}$$

$$< 3\,\frac{e^8 + a^8 - 2e^4a^4}{16} = 3E_4^2 \tag{B.6}$$

Therefore, we have (B.1) as a result of noting (B.6).

Corollary

For the special case where $e = a$, we have from the definition of E_j that $3E_4^2 = 0 = 2E_3E_5$ immediately.

THEOREM B.2

$$E_2E_5 < 2E_3E_4 \quad \text{for } e \neq a \tag{B.7}$$

Proof

$$\frac{30E_2E_5}{e^2 - a^2} = 3(e - a)(e^4 + e^3a + e^2a^2 + ea^3 + a^4) \tag{B.8}$$

$$60E_3E_4(e^2 - a^2) = 5[(e^3 - a^3)(e^2 - a^2)]$$

$$= 5(e - a)(e^2 + ea + a^2)(e^2 + a^2)$$

$$= 5(e - a)[e^4 + e^3a + 2e^2a^2 + ea^3 + a^4] \tag{B.9}$$

Comparing (B.8) and (B.9), we have (B.7) immediately.

Corollary 1

$$5E_2E_5 < 6E_3E_4 \tag{B.10}$$

Proof

Let

$$M = e^4 + e^3a + e^2a^2 + ea^3 + a^4 \tag{B.11}$$

$$N = e^4 + e^3a + 2e^2a^2 + a^3 + a^4 \tag{B.12}$$

Then, we see from (B.11) and (B.12) that

$$M < N \tag{B.13}$$

In fact,

$$N = M + e^2 a^2 \tag{B.14}$$

Also, from (B.8) and (B.11), we have

$$10E_2E_5 = (e^2 - a^2)(e - a)M \tag{B.15}$$

and from (B.9) and (B.12), we obtain

$$12E_3E_4 = (e^2 - a^2)(e - a)N \tag{B.16}$$

Additionally, (B.13) means

$$10E_2E_5 < 12E_3E_4 \tag{B.17}$$

so (B.10) is proved.

Corollary 2

A convenient and interesting way to calculate E_2E_5 and E_3E_4 is via Formulae (B.18) and (B.19) respectively. Thus,

$$E_2E_5 = \frac{(e^2 - a^2)(e - a)M}{10} \tag{B.18}$$

$$E_3E_4 = \frac{(e^2 - a^2)(e - a)N}{12} \tag{B.19}$$

Alternatively, we can relate E_2E_5 and E_3E_4 by P and Q as follows.

$$\frac{E_2E_5}{e - a} = P \tag{B.20}$$

$$\frac{E_3E_4}{e - a} = Q \tag{B.21}$$

where

$$P = \frac{(e^2 - a^2)M}{10} \tag{B.22}$$

$$Q = \frac{(e^2 - a^2)N}{12} \tag{B.23}$$

THEOREM B.3

$$E_3^2 < E_2E_4 \tag{B.24}$$

See Chapter 6 for the proof.

THEOREM B.4

$$3E_3^2 > 2E_2E_4 \tag{B.25}$$

See Chapter 6 for the proof.

Corollary

$$3E_2E_4 > 3E_3^2 > 2E_2E_4 \tag{B.26}$$

Proof

This is a direct result of applying Theorem B.3 and Theorem B.4 simultaneously. In other words, putting Theorem B.3 and Theorem B.4 together literally, we have this corollary.

APPENDIX C.
PROOF OF $S > 0$:
UPPER AND LOWER BOUNDS FOR
THE ENTITY S IN SECTION 6.22

THEOREM

S as defined in Section 6.22, with $e > a$ and $e \neq a$, is positive.

Proof

By the definition given in Section 6.22,

$$S = a_2^2 - 3a_1a_3 = 3Q \tag{C.1}$$

where

$$Q = (12E_3^2k_2^2 - E_2E_4k_1k_3) \tag{C.2}$$

with

$$k_1 = \frac{3}{J_2} + \frac{4}{J_3}, \qquad k_2 = \frac{1}{J_2} + \frac{1}{J_3}, \qquad k_3 = \frac{3}{J_2} + \frac{2}{J_3} \tag{C.3}$$

Claim

$$9k_2^2 > k_1k_3 > 8k_2^2 \tag{C.4}$$

To show that (C.4) is true, we just observe that

$$k_1 k_3 = 9k_2^2 - \frac{1}{J_3^2} \tag{C.5}$$

and

$$k_1 k_3 = 8k_2^2 + \frac{1}{J_2^2} + \frac{2}{J_2 J_3} \tag{C.6}$$

by applying the definitions of k_1, k_2, and k_3 and noting that all the J_i's are positive. As a result of (C.5) and (C.6), we have

$$(12E_3^2 - 9E_2 E_4) < Q < (12E_3^2 - 8E_2 E_4) \tag{C.7}$$

Note that for the proof here, we do not really need the part $k_1 k_3 > 8k_2^2$ in (C.4) or the part $(12E_3^2 - 8E_2 E_4)$ in (C.7), but we will keep them in for completeness and future use regarding the properties of S.

Now consider the entity D, defined as

$$D = 12E_3^2 - 9E_2 E_4 \tag{C.8}$$

Applying the definitions of the E_j's, we obtain

$$D = (e - a)^2 \, \frac{5e^4 + 10e^3 a + 42e^2 a^2 + 10ea^3 + a^4}{24} > 0 \tag{C.9}$$

because $e > a$ and $e \neq a$ from the given premises.

By (C.7) and (C.8), we see that $D < Q$. Invoking (C.9), we see that this means

$$0 < D < Q \tag{C.10}$$

but by definition from (C.1)

$$S = 3Q \tag{C.11}$$

From (C.10) and (C.11), we have $S > 0$. Thus, the theorem is proved.

Next, we will see an upper bound and a lower bound for S by looking at two expressions above. For convenience, let us define

$$U = 12E_3^2 - 8E_2 E_4 \tag{C.12}$$

Then, from (C.1), (C.4), and (C.12), we see immediately that

$$3D < S < 3U \tag{C.13}$$

This means that S has $3D$ and $3U$ as a lower bound and an upper bound respectively.

APPENDIX D.
UPPER AND LOWER BOUNDS
FOR *T* IN SECTION 6.22

The entity T, by the definition given in Section 6.22, is

$$T = a_2^2 - 4a_1a_3 \tag{D.1}$$

which is, after some algebraic operations, equal to

$$T = 4(9E_3^2k_2^2 - E_2E_4k_1k_3) \tag{D.2}$$

where k_1, k_2, and k_3 are as defined in Appendix C.

From (C.4) in Appendix C, we have immediately

$$T > 36k_2^2(E_3^2 - E_2E_4) = L \tag{D.3}$$

and

$$T < 4k_2^2(9E_3^2 - 8E_2E_4) = U \tag{D.4}$$

We see from (D.3) and (D.4) that T has L and U as a lower bound and an upper bound respectively.

Furthermore, U can be rewritten, upon applying the definitions of the E_j's, as

$$U = 4k_2^2(e - a)^2a^2e^2 \tag{D.5}$$

It is interesting to note that $L < 0$ but $U > 0$ and the difference between U and L is

$$D = 4k_2^2(e - a)^2 \frac{B}{8} \tag{D.6}$$

where

$$B = [e^4 + 2ea(e^2 + ea + a^2) + a^4] \tag{D.7}$$

is a nice and neat function of e and a. In fact, (D.7) is a fourth-degree homogeneous polynomial in e and a.

Moreover, B is a positive real-valued function of the positive real-valued e and a, with the following additional properties. For fixed e, entity B is an increasing function of the nonnegative variable a. As a result of this, B has its absolute smallest possible value B_1 within the domain of definition of the variable a when $a = 0$; that is,

$$B_1 = e^4 \tag{D.8}$$

It is interesting to see what happens if $e = a$, which corresponds to a trivial load case when we apply it to beam problems as was done in Chapter 6. Thus, we obtain the other extreme of the value of B, namely

$$B_2 = 8e^4 \tag{D.9}$$

and we see that

$$B_2 = 8B_1 \tag{D.10}$$

What a difference!

APPENDIX E.
UPPER AND LOWER BOUNDS
FOR R_4 IN SECTION 6.23

From the expression for R_4 in Section 6.23, we see that it is expedient to deal with upper and lower bounds for the f_i's first. An upper bound for f_1 is

$$u_1 = 2L_2(L_1 + L_2)\left[\frac{aw}{3b}\right] > 0 \tag{E.1}$$

An upper bound for f_2 is

$$u_2 = -(L_1 + L_2)\left\{3L_2 z_2 + \left[\frac{aw}{3b}\right]\right\} < 0 \tag{E.2}$$

An upper bound for f_3 is

$$u_3 = (L_1 + L_2)\left[\frac{2w}{3b} + 3z_2\right] > 0 \tag{E.3}$$

An upper bound for f_4 is

$$u_4 = \frac{-(L_1 + L_2)w}{3bL_2} < 0 \tag{E.4}$$

Hence, an upper bound for R_4 is

$$U = k'[u_1 E_2 + u_2 E_3 + u_3 E_4 + u_4 E_5] \tag{E.5}$$

Next, we do a lower bound for R_4 via lower bounds for the f_i's, as follows. A lower bound for f_1 is

$$n_1 = \frac{L_2(L_1 + L_2)aw}{2b} > 0 \tag{E.6}$$

A lower bound for f_2 is

$$n_2 = -(L_1 + L_2)\left[4L_2z_2 + \frac{aw}{2b}\right] < 0 \tag{E.7}$$

A lower bound for f_3 is

$$n_3 = (L_1 + L_2)\left[\frac{w}{2b} + 2z_2\right] > 0 \tag{E.8}$$

A lower bound for f_4 is

$$n_4 = -(L_1 + L_2)\left[\frac{w}{2bL_2}\right] < 0 \tag{E.9}$$

Hence, a lower bound for R_4 is

$$N = k'[n_1E_2 + n_2E_3 + n_3E_4 + n_4E_5] \tag{E.10}$$

Note that z_2 is used in the above in the expressions for u_3 and n_3 purely for brevity, where z_2 stands for

$$z_2 = \left(1 + \frac{a}{L_2}\right)\frac{w}{6b} \tag{E.11}$$

which is also Formula (6.1103) in Section 6.23.

APPENDIX F.
SINGLE-SPAN BEAMS WITH A FIXED END
AND A SIMPLY SUPPORTED END

UNIFORM LOAD

Refer to Figure F.1. The reactions are

$$R_1 = wb \left(\frac{\dfrac{3D}{2} + L^3 - d^3}{L^3} \right) \tag{F.1}$$

$$R_2 = wb - R_1 \tag{F.2}$$

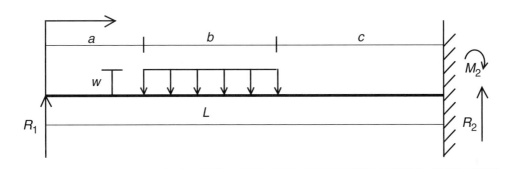

FIGURE F.1. Uniform load on a single-span beam with a fixed end and a simply supported end

where

$$D = b^2 \frac{4a + 3b}{12} - (L^2 - d^2) \frac{a + d}{2}, \qquad d = a + b \qquad (F.3)$$

The end moments are

$$M_1 = 0 \qquad (F.4)$$

$$M_2 = R_1 L - wb\left[L - \left(a + \frac{b}{2} \right) \right] \qquad (F.5)$$

The deflections are

$$Y = \left[\frac{1}{EI} \right]\left[Fx - \frac{R_1 x_3}{6} \right] \quad \text{for } 0 < x < a \qquad (F.6)$$

$$Y = \left[\frac{1}{EI} \right]\left[Fa - \frac{R_1 a^3}{6} \right] \quad \text{at } x = a \qquad (F.7)$$

$$Y = \left[\frac{1}{EI} \right]\left[Fx - \frac{R_1 x^3}{6} + \frac{w(x - a)^4}{24} \right] \quad \text{for } a < x < a + b \qquad (F.8)$$

$$Y = \left[\frac{1}{EI} \right]\left[(a + b)F - R_1 \frac{(a + b)^3}{6} + \frac{wb^4}{24} \right] \quad \text{at } x = a + b \qquad (F.9)$$

$$Y = \left[\frac{1}{EI} \right]\{ 3[R_1 L - wb(L - e)] + (wb - R_1)z \} \frac{z^2}{6} \qquad (F.10)$$

$$\text{for } a + b < x < L$$

where

$$F = \frac{R_1 L^2}{2} - wb \frac{3c(b + c) + b^2}{6} \qquad (F.11)$$

$$z = L - x, \qquad e = a + \frac{b}{2} \qquad (F.12)$$

Some important special cases are as follows.

Case 1. a = 0

The reactions are given by (F.1) and (F.2), with

$$D = b \frac{3b^2 - 2L^2}{4} \qquad (F.13)$$

The end moments are given by (F.4) and (F.5), with $a = 0$. The deflections are given by (F.8–F.12), with $a = 0$.

Case 2. $c = 0$ and $d = L = a + b$

The reactions are given by (F.1) and (F.2), with

$$D = b^2 \frac{4a + 3b}{12} \tag{F.14}$$

The end moments are given by (F.4) and (F.5). The deflections are given by (F.6–F.9) and (F.11), with $c = 0$.

Case 3. $a = 0$ and $c = 0$

The reactions are

$$R_1 = \frac{3wL}{8} \tag{F.15}$$

$$R_2 = \frac{5wL}{8} \tag{F.16}$$

The end moments are given by (F.4) and (F.5), with $a = 0 = c$ and $b = L$. The deflections are given by (F.8), (F.9), and (F.11), with $a = 0 = c$ and $b = L$.

TRIANGULAR LOAD

Refer to Figure F.2. For the general case with nonzero a and c, the results are as follows. The reactions are R_1 and R_2, with

$$R_1 = A + B + D \tag{F.17}$$

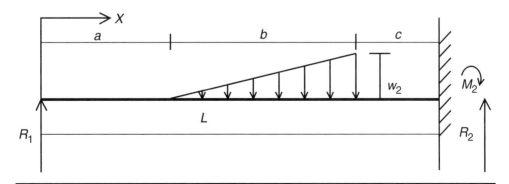

FIGURE F.2. Triangular load on a single-span beam with a fixed end and a simply supported end

where

$$A = w\left(\frac{b}{L}\right)^3 \frac{5a + 4b}{40}, \qquad B = wb\,\frac{1 - \left(\dfrac{d}{L}\right)^3}{2} \tag{F.17A}$$

$$D = -wb(3a + 2b)\,\frac{1 - \left(\dfrac{d}{L}\right)^2}{4L}, \qquad d = a + b \tag{F.17B}$$

$$R_2 = \frac{wb}{2} - R_1 \tag{F.18}$$

The end moments are

$$M_1 = 0 \tag{F.19}$$

$$M_2 = R_1 L - wb\,\frac{L - \left(a + \dfrac{2b}{3}\right)}{2} \tag{F.20}$$

The deflections are

$$Y = \left[\frac{1}{EI}\right]\left[Fx - \frac{R_1 x^3}{6}\right] \quad \text{for } 0 < x < a \tag{F.21}$$

$$Y = \left[\frac{1}{EI}\right]\left[Fa - \frac{R_1 a^3}{6}\right] \quad \text{at } x = a \tag{F.22}$$

$$Y = \left[\frac{1}{EI}\right]\left[Fx - \frac{R_1 x^3}{6} - w\,\frac{(x - a)^5}{120b}\right] \quad \text{for } a < x < a + b \tag{F.23}$$

$$Y = \left[\frac{1}{EI}\right]\left[F(a + b) - R_1\frac{(a + b)^3}{6} - \frac{wb^4}{120}\right] \quad \text{at } x = a + b \tag{F.24}$$

$$Y = \left[\frac{1}{EI}\right]\left[(wb - 2R_1)\frac{z^3}{12} + (2R_1 L - wbc)\frac{z^2}{4}\right] \tag{F.25}$$
$$\text{for } a + b < x < L$$

where

$$z = L - x \tag{F.26}$$

$$F = \frac{R_1 L^2}{2} - \frac{wb^3}{24} - wbc\,\frac{2b + 3c}{12} \tag{F.27}$$

Some important special cases are as follows.

Case 1. a = 0

The reactions are given by (F.17) and (F.18), with simplified A, B, and D due to $a = 0$. The end moments are given by (F.19) and (F.20), with $a = 0$. The deflections are given by (F.23–F.27), with $a = 0$ and $L = b + c$.

Case 2. c = 0 and L = a + b

The reactions are given by (F.17) and (F.18), with $B = 0$ and $D = 0$ due to $c = 0$ and $L = a + b$. The end moments are given by (F.19) and (F.20), with $c = 0$. The deflections are given by (F.21–F.23) and (F.27), with $c = 0$.

Case 3. a = 0, c = 0, and b = L

The reactions are

$$R_1 = \frac{wL}{10} \tag{F.28}$$

$$R_2 = \frac{2wL}{5} \tag{F.29}$$

The end moments are

$$M_1 = 0 \tag{F.30}$$

$$M_2 = \frac{-wL^2}{15} \tag{F.31}$$

The deflections are

$$Y = \left[\frac{1}{EI}\right]\left[Fx - \frac{R_1 x^3}{6} - \frac{wx^5}{120b}\right] \quad \text{for } 0 < x < L \tag{F.32}$$

with

$$\max Y = \left[\frac{1}{EI}\right]\left[F - \frac{R_1 u}{6} - \frac{wu^2}{120b}\right]u^{1/2} \quad \text{at } x = +u^{1/2} \tag{F.33}$$

where

$$F = \frac{R_1 L^2}{2} - \frac{wb^3}{24} \tag{F.34}$$

$$u = \frac{12b}{w}\left\{\left[\frac{R_1^2}{4} + \frac{wF}{6b}\right]^{1/2} - \frac{R_1}{2}\right\} \tag{F.35}$$

This appendix touches upon the treatment of a very simple type of statically indeterminate beam. Exploration is always beneficial, and there are at least two tasks that the interested reader can do to further explore this topic. One task is to write out in detail, according to the guidelines given here, all the formulae mentioned for the special cases of importance and compare the results among the cases. The other is to set up a project using the expression for R_1 (or R_2) to investigate the effects of a selected parameter on R_1 (or R_2).

INDEX